Coral Reefs: A Very Short Introduction

VERY SHORT INTRODUCTIONS are for anyone wanting a stimulating and accessible way into a new subject. They are written by experts, and have been translated into more than 45 different languages.

The series began in 1995, and now covers a wide variety of topics in every discipline. The VSI library currently contains over 650 volumes—a Very Short Introduction to everything from Psychology and Philosophy of Science to American History and Relativity—and continues to grow in every subject area.

Very Short Introductions available now:

Available soon:

For more information visit our website

www.oup.com/vsi/

Charles Sheppard

CORAL REEFS

A Very Short Introduction

SECOND EDITION

OXFORD
UNIVERSITY PRESS

OXFORD
UNIVERSITY PRESS

Great Clarendon Street, Oxford, OX2 6DP,
United Kingdom

Oxford University Press is a department of the University of Oxford.
It furthers the University's objective of excellence in research, scholarship,
and education by publishing worldwide. Oxford is a registered trade mark of
Oxford University Press in the UK and in certain other countries

First edition published 2014
This edition published 2021

Impression: 1

Published in the United States of America by Oxford University Press
198 Madison Avenue, New York, NY 10016, United States of America

British Library Cataloguing in Publication Data
Data available

Library of Congress Control Number: 2020947755

ISBN 978-0-19-886982-5

Printed and bound by
CPI Group (UK) Ltd, Croydon, CR0 4YY

Contents

List of illustrations

Underwater photographs by Anne Sheppard and Charles Sheppard.

Chapter 1
Geology or biology?

Discovery and early explorers

Coral reefs have been regarded with awe for centuries. Early seafarers were wary of them, naturalists were confused by them, but many coastal peoples benefited greatly from these mysterious rocky structures that grew up to the surface of the sea. They have been the source of food for coastal peoples, and they provide a breakwater from storms and high waves to countless coastal communities that developed in their lee.

Long before scientists wondered what coral reefs really were, they had been encountered and used in many ways. Awareness of reefs beyond the coastal villages that used them for food started, perhaps, amongst pilgrims sailing down the Red Sea to Jeddah, the port serving the holy city of Mecca, for that part of the Red Sea is one area that is gloriously rich in—or dangerously strewn, from a seafarer's point of view, with—coral reefs. And then, from a European perspective, as trade developed to both the East and West Indies, coral reefs were increasingly encountered—as navigational hazards that provided fatal resting places for countless traders. This was because their biology leads them to grow up to the surface of the sea at low tide, just where they offer the greatest danger to sailing ships. Perhaps the most famous European explorer to suffer the consequences was Captain Cook, who nearly came to an early demise on the Great Barrier Reef as he explored the coast of Australia.

Nevertheless, the value of coral reefs has always far exceeded the hazards they pose to navigation. Their distribution is tropical and widespread, and ever since humankind spread out from Africa, migrating eastwards through the Indonesian and Philippine archipelagos and onwards to the Pacific Islands, people have inhabited land near reefs, attracted by the huge quantities of fish and invertebrates available for food. People living beside a coral coast commonly obtained almost all their protein from fishing or gleaning at low tide, and many still do. Even those who lived a little further from the water would obtain a significant proportion of their food from those who caught more from the reef than they needed.

Europeans mostly travelled west across the Atlantic at first, where they encountered the coral fringed islands of the Caribbean. Columbus's first New World landfall was made tricky by the coral reefs of San Salvador island in the Bahamas (some say), which the intrepid navigator thought was Cathay on the other side of the world. Following the earliest days of exploration came the age of naturalists. At this time, the coral island of Bermuda, which lies to the north of the main Caribbean region, became important in providing these early exploring naturalists with a half-way stop on their journey. This period of exploration promoted the beginnings of modern scientific understanding, because many of the ships of discovery also carried a naturalist as a passenger, possibly appointed by a king or an archduke on whose patronages so much early discovery depended. Slightly later, ships bearing these naturalists headed eastwards too, sometimes on journeys of exploration, but more commonly as appointed passengers. One naturalist, Carsten Niebuhr, wrote of coral reefs as he travelled east:

I have already had occasion to speak, in the course of my travels, of the astonishing mats of works formed by the marine insects; namely the immense banks of coral bordering, and almost filling up [the sea]. The reader may therefore conceive with himself what a variety of madrepores and millepores are to be met with in these seas.

He wrote this in 1792. There was in those days, and for a long time to come, an almost complete inability to see or understand what went on beneath the surface of the water. Sea monsters still decorated nautical charts, and the monsters were only slightly less useful than the positions marked of many of the reefs. Fifty years later, in 1838, also travelling eastwards, J. Raymond Wellstead was still mystified by it all:

there are secrets on the surface as well as within the bosom of the ocean, which lie shrouded from human observation and research...where there is mystery there will always be interest, and to the greater the one, the more intense the other.

Another naturalist of that time, C. G. Ehrenberg, who became very well known in coral science circles, wrote in 1832:

hummingbirds sport around the plants of the tropics, so also small fishes, scarcely an inch in length and never growing larger, but resplendent with gold, silver, purple and azure, sport around the flower-like corals.

These naturalists were clearly fascinated by coral reefs and were frustrated by not being able to see them properly. But some were starting to get closer to them; another half-century later, in 1878, C. B. Klunzinger wrote:

Nowhere can one contemplate the life of the corals, and what belongs to it, more quietly and comfortably than here, although he has to lie on his belly—a trifling matter for the naturalist—and hold his magnifying glass at the point of his nose above a coral bush.

Meanwhile Alfred Russel Wallace, the co-discoverer of evolution, wrote of Indonesian reefs in 1869:

> The clearness of the water afforded me one of the most astonishing and beautiful sites I have ever beheld. The bottom was absolutely hidden by a continuous series of corals, sponges, actinae, and other marine productions, of magnificent dimensions, varied forms and brilliant colours … it was a sight to gaze at for hours, and no description can do justice to its surprising beauty and interest.

And this amazement came after he had explored such exotic parts of the world as the Amazon, not to mention Indonesia's islands.

Explanation of how reefs grow—the 'Coral Reef Question'

In the 19th century geologists led research on coral reefs, building an understanding of how these remarkable structures developed. Considerable controversy and impassioned arguments grew around how reef structures grew upwards to sea level and then stopped. In one sense this was easy: they were somehow produced by marine life which would not grow out of the water. But what were they—animals or plants? And how did they make the rock? The term 'animalcule' was coined at one stage to describe the baffling small bunches of waving tentacles that were attached to the top of the rocks they sat upon, some a few inches long, but mostly tiny. It was clearly very confusing at the biological level, though at the geological level the early scientists were starting to see what had actually happened.

Scientists knew by then that this rock was limestone, calcium carbonate, and a very pure form of it. They knew too that the shells of many animals were also made of limestone. But although it was well known that marine organisms produce limestone, these reef structures were immense. Inland, many miles from the sea, were other limestone hills and mountains, clearly very ancient. Could marine organisms have built these too? They certainly contained marine fossils, which suggested that indeed they had—that is, unless the naturalist had a biblical view of time spans, which certainly posed an unsolvable problem to quite a number of these naturalists. Before the full recognition that the Earth was ancient, the implied vast timescales served to baffle as much as to enlighten.

Charles Darwin's famous voyage on the *Beagle* was one of the turning points in coral reef science. Although better known for his ideas on evolution, I think his 1842 book on coral reef formation and large-scale movements of parts of the Earth's surface was equally innovative and original. His theory depended on the massive subsidence of volcanoes (and other parts of the Earth's surface). Darwin proposed that coral animals could grow only near the surface and, as the land against which they grew subsided, the corals kept growing upwards, on top of each other, to maintain their position in shallow and warm waters. It had already been observed, in sections taken from fossil reefs, that corals were embedded in the limestone matrix, so this was not the shock; what surprised and was contested by some was the huge timescale needed, if Darwin's explanation was correct. His theory had a simple but powerful elegance along with its explanatory power, and it eventually became accepted and then was proved long after his own lifetime.

But in Darwin's time this was still only a partial theory, and it did not displace all the earlier theories for many years. For example, in 1821 Von Chamisso had noticed that reef-building corals grow best in turbulent waters that are found on the outer periphery of most reefs. This led him to believe that in such conditions corals would grow upwards so that, whatever their foundation, they would reach the surface in the annular shape typical of a coral atoll. Two famous collaborators, Quoy and Gaimard, opined that corals simply grew up around the rim of old volcanoes. The coincidence of so many volcanoes happening to grow to identical height just below the water surface was never explained. Nevertheless, the latter theory was even supported by the great geologist Charles Lyell until he dropped it in favour of Darwin's theory.

The accepted mechanism today is this: before any subsidence event, for example around the flanks of a new volcano—anywhere on the flank, not specifically the rim—corals would establish in the photic zone (i.e. where there is enough light). The volcano would then sink a little and corals would grow up, fringing the island and forming what we call a fringing reef (Figure 1). Then, with further subsidence of the volcano and upward growth of the reef, the reef appears (from the surface) to detach from the land and to form a reef that is increasingly distant from shore—a barrier reef. One definition of a barrier reef has been simply that there should be a navigable passage between the reef and the shoreline.

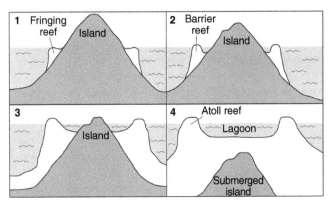

1. Sketch of Darwin's scheme of the sequence of progression of a fringing reef, barrier reef, and atoll as its central volcano subsides over millions of years.

The final stage of this sequence is that further subsidence submerges the volcanic island completely so that all that remains visible from the surface is a ring of coral: the coral atoll. It is not the case that a ring of coral is simply sitting on top of the rim of an old volcano—this is far too simplistic. For one thing the old volcano today might be hundreds or thousands of metres deeper, and we now also know that repeated lowering and rising of sea level with glacial cycles has greatly changed and modified any such simple structure, and that there have been repeated episodes of growth interlaced with periods of no growth and erosion. Also, limestone is relatively easily dissolved by rainwater which means that substantial erosion of old reef above sea level also occurred when sea level dropped in one or other of several geological periods. All these add to the complexity of what we see at the surface today.

Darwin's broad theory of atoll formation by subsidence is generic and, while accurate enough, it only partially explains the formation of so many of the world's complicated networks of barriers, islands, and fringing reefs. The many exceptions and complications to Darwin's simple progression led to more than a hundred years of further debate, showing that this was a complex problem. Coral reefs will grow upwards to meet the surface when sea level is high, and will be eroded away during periods when sea level is low. Progressive understanding not only of vertical land movement but of vast changes in sea level lasting thousands of years, usually during ice ages, showed the layers of complexity that superimpose Darwin's central, but basically correct, idea.

Complexity and proof

Numerous drilling projects attempted to penetrate through reefs to reach bedrock in various places and prove Darwin's theory, such as at Funafuti atoll in 1898, on the Great Barrier Reef in the 1920s and 1930s, and in Borneo in the 1940s, but none of these reached any material other than coral rock for the several hundred metres drilled before the projects were abandoned. Finally, a deep boring to 1,340 metres in Bikini atoll in the Pacific struck underlying volcanic foundation. However, it is also recognized that sea level changed. Reginald Daly's Glacial Control hypothesis, developed in the early part of the 20th century, showed how great changes in sea level were crucial to reef formation too. At the time, many viewed it as a challenge to Darwin's subsidence theory, but in reality, it simply complements it. David Stoddart wrote: 'If Darwin's was a theory of reef structure, Daly's was one of reef surface morphology.' Erosion has had a great effect on the present topography of reefs, particularly in areas of the world with relatively rapid land movements.

Several other ideas tried to explain differences seen in various places. The antecedent platform theory was developed to explain why barrier reefs and atolls can develop on any suitable submerged platform lit by the sun, and that fringing reefs need not form an intermediate stage of the process. This theory required vast numbers of similarly located platforms, and an explanation for this unlikely event included wave erosion, which might reduce all volcanoes to the same shallow depth. The theory emphasized the importance of erosion, making it as important as the growth of corals. It was known, too, that erosion of old reef by rain results in saucer-shaped depressions in large areas of limestone, which creates atoll-like dishes. This karstic-saucer theory developed by Edward Purdy certainly explains many structures within reefs, such as the deep cavities in reef rock known as 'blue holes'.

Scientists often tend to have a preference for one single theory, but different theories may be plausible, depending on place, scenario, and time. This is still true now in relation to what is causing today's demise of coral reefs generally. Scientists may be strongly influenced by where in the world they have worked and which reefs they have observed. In fact, all these processes take place to a greater or lesser degree in different places, but the possibility of a scientist visiting sufficiently numerous and varied locations to see all possible examples of reefs the world has to offer is difficult even today, and was simply not possible in the days of sailing ships.

Biological discovery

While these geological explanations were being developed, the reef's component corals, fish, and other groups were being described extensively by scientists (Figure 2), though perhaps with limited understanding. Not until about three-quarters of a century after the early geologists' explanations did biologists start to understand life on the reefs in an ecological way, to discover how the reefs worked, how their components fitted together, and what their most important components were. The colours and apparent profligacy of life on reefs had long attracted naturalists who were

2. Early taxonomic drawings of corals were very accurate and detailed, though they were usually done from fragments that were dredged up and sometimes rather damaged in the process.

much more used to life on the shores of their own, usually colder, countries. However, an ecological understanding of function awaited the advent of diving apparatus, even if some important progress had been made through the use of microscopes and experiments in laboratories and on very shallow reefs.

This book is about the biology and ecology of reefs but, as reefs lay down vast volumes of limestone, it includes some geology too. This then leads to description of events that have overwhelmed many of the world's reefs during the past few years. These events are largely a result of climate change, mainly warming pulses that are related to globally important El Niño events. The obvious biological complexity of reefs makes them rather daunting to understand. Reefs occupy about 1 per cent of the ocean's area yet support about 25 per cent of marine diversity, and reefs themselves are built by that life. Their complexity is matched by their importance: they show enormously high rates of protein production—when they are not damaged by too much exploitation. This being the case, they are increasingly important to the growing number of people who demand their 'services'—as economists put it. Coral reefs are now estimated to provide a value, in monetary terms, of more than a third of a million dollars per hectare, per year, made up from food supply to local people, from tourism for visitors who require food, accommodation, transport, and recreational services, but also from their breakwater effect, something that is becoming increasingly recognized. Healthy and growing reefs protect shorelines wherever they are located, from tropical villages to high value coastal cities, infrastructure, factories, and other developments. A biologist might not much like reducing the idea of a reef to something that simply provides food, tourism, and shelter, but in this overcrowded and overexploited world that is increasingly the way of society. And there is more to a reef's value too, which lies in aspects such as provision of huge biodiversity and simple beauty, neither of which is usually quantified.

To many scientists, reefs are the most intriguing ecosystems that have ever appeared in the world's oceans. They contain many scientific paradoxes: in some ways they are robust, but in others they are very fragile. They contain a riot of species which appear to be visually chaotic, but the organization behind the apparent chaos is being better understood all the time. Coral reefs provide material for greater insights into how nature is organized. Hopefully we will find a way to interact with and use reefs with a minimum of disruptive impact. We should even be able to please economists and governments by showing how reefs provide services to the burgeoning human population.

Chapter 2
Ancient reefs and islands

Ancient reefs

We should not think of coral reefs as being a permanent fixture of the Earth. In the past, several other groups of organisms formed reefs for hundreds of millions of years but then became extinct. There are many ancient reef formations built from a wide range of organisms that built massive limestone structures on our planet, uplifted to form hills and mountains.

In Precambrian times, assemblages of simple marine organisms laid down reefs over aeons of time. These were stromatolites, built by photosynthetic cyanobacteria, depositing limestone but also trapping sediment to form layered domes. In that Archaean Eon, long before the advent of complex life, atmospheric oxygen was low.

Atmospheric oxygen increased significantly from about 1.8 billion years ago, a change that permitted the development of more complex, multicellular, oxygen-using forms of life. Existing reefs declined but around the beginning of the Cambrian Period a new group of reef-building organisms called archaeocyathids, small sponge-like organisms that were probably filter feeders, developed to reach their peak. This period saw an explosion of animal life. The archaeocyathids became extinct at the end of the Cambrian when, for a few more tens of millions of years, stromatolites were again the main reef-builders. Around that time sponges, distant ancestors of some of today's sponges, became important reef-builders. Sponges never went away, and are still abundant on reefs. These were joined for many millions of years by stromatoporoids, which appear to be related to sponges, and which dominated from the Ordovician Period but began to decline in the Carboniferous—the time, some 350–300 million years ago, during which thrived the lush vegetation whose remains eventually became transformed into today's great seams of coal.

Two groups of corals then began to form reefs, but these were not the ancestors of today's corals. The first group was the tabulate corals, which were almost entirely colonial, and the rugose corals which existed in colonial and solitary forms. Both groups lasted until the end of the Permian (about 250 million years ago), and during this long span they built some extensive reefs. All this came to an end at the great Permo-Triassic extinction event. This extinction had a massive effect on the world's life, seeing the demise of perhaps over 90 per cent of marine species. As Raup put it in 1998: 'if these estimates are even reasonably accurate, global biology (for higher organisms at least) had an extremely close brush with total destruction'. Even sponges, which thrive today, came close to extinction. Several causes have been suggested for this extinction, but a likely possibility is that the climate changed, a salutary point today.

The main reef-building animals of today, the scleractinian corals, came to dominance about 240 million years ago. These are cnidarians, but what exactly they developed from is debated; they may have arisen from anemone-like relatives, or may be ancestral to them, and possibly different groups or families of corals may have different origins. Corals do not appear in the fossil record until well after the great Permian-Triassic extinction event, although that geological interval has been very well investigated.

Their biology is described later, but they became so successful as reef-builders because they can form symbiotic associations with single-celled algae called zooxanthellae, so that by the middle Triassic they were forming substantial limestone structures. This symbiosis is a crucial aspect of coral reef life.

The next great extinction occurred at the end of the Cretaceous Period. The so-called K-T extinction, which removed the dinosaurs, killed about 70 per cent of corals, and greatly suppressed reef-building for a long time, but corals then returned with vigour.

The deposition of limestone has been a characteristic of all these reef-forming groups. By doing so they could both create their own substrate and modify their environment to a considerable extent. The deposition of limestone requires energy, and certain conditions need to be met in terms of carbon dioxide–carbonate balance, temperature, acidity and other physical factors. Thus the growth of these structures requires fairly tight environmental parameters. Since the earliest days of corals, atmospheric and oceanographic conditions on Earth have varied, often substantially, but corals have thrived throughout this time, forming the antecedents for today's coral reefs.

This history is important. Present-day ecology—basically, how a reef works—is a product of this long ancestry. Climate has fluctuated greatly throughout this time; sea levels have changed, and the temperatures of the Earth have changed too. The riot of life seen on healthy reefs today is the result of a very long heritage.

Coral islands—tips of the icebergs

Coral islands have long been a key component of coral reefs and the target of research. They are commonly the first parts of reefs that are seen. They are made from reefs, and sit on top of them.

Home to millions of people, they come in different forms, but in terms of their structure there are really only a few different kinds, with many islands showing combinations of all of them. The simplest are the true coral cays. These range from being little more than a sand bank to larger islands, richly clothed with vegetation. They may form when sand, itself a product of a broken down and eroded coral and reef matrix, gets pushed onto the top of a reef by wave action. If enough sand is deposited, birds may settle. The birds leave guano, which in turn fertilizes seeds that have arrived by wind and birds; some plants take root; and this attracts more birds, who leave more guano. As more sand gets pumped onto this now dry platform, plant roots stabilize it and prevent it from being immediately washed away in a storm. In time a coral cay is born. Then, with help from microbes and various chemical reactions, the sand grains become cemented together to form solid rock. Details of how this happens remain poorly understood, but the result is a small island (Figure 3).

A more complicated variant in coral island formation contains cliffs that rise several metres above sea level. These islands have solid blocks of limestone at their core, which are made up of much older reefs or limestone sand dunes, formed when sea levels were higher than they are now, before or during the last ice age. At the maximum extent of glaciers these blocks protruded above present sea level by many tens of metres, and indeed tall limestone stacks can be seen protruding from the sea today in several parts of the world. A gradual rise of sea level starting about 18,000 years ago, as well as erosion of the limestone rock from rain, has left emergent lumps of rock protruding just a few or a few tens of metres above the surface of the sea. The islands' cores may be a mixture of consolidated limestone rubble and rock. They may have shifting areas of sand around the edges, but their altitude and solid core gives away the fact that the hearts of these islands are of much older limestone. Carbon dating shows that they can

3. Three small islands in the central Indian Ocean. All have an old core of limestone 8,000 years old, the larger two are covered with recent sand and rubble. The foreground island has an extensive and complicated reef flat, much less reef flat surrounds the furthest one. Note the bite in the bottom right of the nearest island, where rising sea levels and reef mortality are eroding the land.

be anywhere between 6,000 and several hundred thousand years old. Many islands in the Caribbean are of this kind.

Then there are 'high islands', which are basically volcanic or basaltic islands fringed with coral reef. Many islands in the Pacific and Indian Oceans, and large numbers in the Caribbean are of this kind—most are ancient or even currently active volcanoes, or other land forms which may still be subsiding into the sea. They may have lush interiors, if they benefit from high rainfall, or they may be more barren where prevailing winds do not bear much moisture. One can see all variants of coral island in many countries, often all within one visual sweep of the landscape.

The coralline or reefal component of some islands may be relatively small compared to other rock.

The result is countless coral islands around the tropical world. Many entire nations are built exclusively on coral reef islands and it is salutary to remember that a significant number of states in the United Nations General Assembly are there only because of corals and reef processes. As we will see later, many of these countries are in trouble because of the problems arising from being only marginally above sea level, and being dependent on the living processes of the reefs. This makes them potentially very vulnerable. Rain is a great structuring force on these islands, both in its amount and its frequency. In those parts of the ocean where there is very little rain, coral islands may never have supported very much life, and may never have been successfully colonized or developed. The substrate after all is very porous and fresh water may not remain in the soil for very long. In other places, heavy rainfall may last for several months each year and much fresh water penetrates the soil and porous rock of the island, forming lenses beneath the land surface, slightly above the level of the sea. It is this water table that keeps seawater at bay, and which supports island vegetation. Where rain is plentiful, island life will thrive, commonly those species that show some tolerance to salt. Where this is the case, human communities inevitably have developed. Many have become an attractive destination today for tourists too; reef-based tourism in some coral island nations can form 40 to over 80 per cent of their national income.

Today's reef crisis

But we are changing the environment around these coral reefs, which is having severe consequences. The present changes to ocean temperature and acidity mean that, at current rates of change, the present rate of coral reef formation simply cannot continue for much longer. Increasing numbers of scientists have given increasingly urgent warnings, and coral reefs as a whole have been described as being the canaries in the coal mine—they are the part of the ecosystem that will be the first to succumb to overexploitation and abuse.

Since direct observations of reefs below a couple of metres' depth started with the introduction of diving equipment, we have had little more than sixty years of modern coral reef research, which has allowed us to further our understanding of these remarkable systems. During this time, several thousand reef researchers have advanced our understanding of how these structures function. While this appears to be a large number of researchers, it is in fact no more than the number that may be found at a single medical or business conference at a single time. Neither is it such an impressive number when we consider that hundreds of millions of people in the world depend on reefs for food, land, or revenue. Unfortunately, and based on present trends, this focus of biological research might be short-lived. In this time, marine scientists have frequently observed, worried, and issued warnings about many aspects of degradation of coral reefs in all our oceans, especially those in some of the more densely populated and heavily utilized areas.

Reefs today, so soon after the start of this 'modern' period of scientific flourishing, are dying. Over a quarter have already died and another half have degraded. Yet although scientists' warnings have grown increasingly grim, those in authority have continued to ignore them. Take the Caribbean, for example. In the 1970s the shallowest 3 metres of just about all reefs there were covered with the world's largest species of coral, *Acropora palmata*, known as elkhorn coral. At low tide you simply could not cross over such a reef because of the presence of these dense forests of branches that grow as tall as a person. Spectacular in the extreme, no diving magazine would miss an opportunity to print new photos of them. But, by the 1980s something called White Band disease had killed almost all of them, a disease linked to human sewage. These huge coral skeletons crumbled, so that by the 1990s most of these glorious forests were rubble. The once three-dimensional reef structures lost their life and no longer broke the waves or protected shorelines. Nowhere are there high densities of elkhorn remaining—the entire shallowest portion of the Caribbean reef ecosystem has essentially disappeared. It is the same story elsewhere: in the coral-rich Gulf of Aqaba I used to observe uncountable numbers of table corals, also of the genus *Acropora*, but, during a visit to Jordan twenty years later, I was taken, proudly and touchingly, to view the only table coral thought to be left in Jordan at that time. Similarly, in the Persian Gulf, some students showed me where coral reefs in the 1960s had vividly first demonstrated that corals could live in salty seas, but we saw not one left alive there in two hours of dives.

What caused these mortalities, which happened so soon after the start of the great age of coral reef discovery? That is explained later but, briefly, reefs have been killed by two groups of impacts. Some are 'local factors', which include impacts from sewage and industrial pollution, along with shoreline dredging, construction, and overfishing. Others are 'global factors' which are all related to the rising amount of CO_2 in the atmosphere, which causes temperatures to rise and the seawater to become less alkaline (acidification).

For those who saw these reef ecosystems before their degradation, or for those fortunate to see some of the remaining reefs in truly good condition, the decline witnessed in a single observer's lifetime has been as scientifically spectacular as it has been alarming. Coral reefs offer us an important warning. We have been ignoring the health and welfare of an ecosystem that supports us, one which has brought pleasure, food, and value to so many. They offer us a mirror, or a microcosm perhaps, of humanity as a whole.

Chapter 3
The architects of a reef

A diver seeing his or her first coral reef may be visually overwhelmed. Hundreds of coral species and thousands of colourful fishes and other species confuse the eye. Fish add tremendous activity, feeding on bottom-dwelling life and on each other (Figure 4). Deeper, sea fans as tall as the diver flex as each wave passes overhead, and crackling noises made by reef creatures fill the sea with sound.

There is structure to this kaleidoscope of course, acting at many scales. There is a pattern across the world, a pattern across a single reef (with many similarities and consistencies between most reefs), and a pattern in the small area that a diver can see at any one time.

The global pattern

Coral reefs are tropical ecosystems. While all reefs contain a very high diversity of life, different areas of the coral seas support large differences in diversity of animals and plants. Southeast Asia is especially diverse while some areas such as the Caribbean and Eastern Pacific have comparatively low diversity (Figure 5). However, reefs in these last two areas are every bit as substantial as those in higher diversity areas because high diversity does not necessarily equate with good or strong reef construction. Indeed,

4. Thriving reef slopes in a no-fishing marine reserve. Top: fish frequently dominate the scene. Bottom: fields of tall sea fans in deeper water.

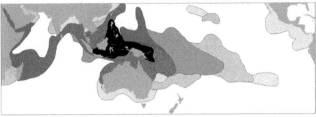

Species of corals

<50   50–100   100–200   200–300   300–400   >400

5. **Map showing the distribution of reef-building corals and their diversity. The 'coral triangle' is in black, in the Southeast Asian region.**

in some diverse areas, corals may be unable to create biogenic reefs at all; typically, the corals may grow on a rock, usually igneous, from which they readily detach after death without any build-up of limestone, yet the associated seabed and fish fauna will be those of a coral reef.

The high diversity of Southeast Asia and the steady decline in diversity as one moves further away from it have long been recognized but are still not well explained. Diversity broadly decreases eastwards and westwards, and there is also a marked decline in diversity of most taxonomic groups along a latitudinal gradient, being highest nearest to the equator and falling towards cooler water. This is seen very clearly in the decline in coral diversity as one moves southwards along the Great Barrier Reef, from well over 300 species in the northern and central regions to 244 in the south, and just 87 at Lord Howe Island at 32° South. Similarly, over 50 species of corals can be found in the Caribbean, but only 21 are known from Bermuda which, at 32° North, is the most northerly reef system. Importantly however, the vibrancy and productivity of a reef appears to be more or less unrelated to its diversity.

Set within this global pattern are localized patterns resulting from regional environmental characteristics. Some regions may be particularly rich, for example the Red Sea, which appears to trap larvae from the Indian Ocean as well as having many endemic species. In other regions coral species may struggle to grow because of factors such as increased amounts of dissolved nutrients derived from river discharges or oceanic upwelling systems. Salinity and temperature extremes constrain corals as does the remoteness of the region and the corals' position relative to prevailing currents that bring larvae from richer areas. The reef region with the poorest diversity is that of Brazil, where reefs contain just nineteen coral species because of its isolation from the Caribbean due to the massive freshwater barrier caused by the Amazon and Orinoco rivers; nearly half of Brazil's corals are endemic too. Tropical West Africa has some corals, but these fail to form reefs at all because of the highly turbid waters there.

Coral animals—the dominant architects

Corals are usually the dominant biota on coral reefs. They are animals, mostly living in colonies but with some solitary species. All lay down a skeleton of limestone. The proportion of each species or group depends, amongst other things, on water quality, exposure, and depth of the site.

The world's coral reefs are not a single, homogeneous ecosystem. Think of forests of trees: there are pine forests, oak forests, beech forests, and many others, each containing different dominant and structural species; some are virtually monocultures (like a pine forest) and some have much higher diversity (e.g. a tropical rainforest). With reefs too, there are many kinds formed by different combinations of species, and the main architectural forms in each case can differ markedly. In the Indo-Pacific, branching forms of the genus *Acropora* dominate in many areas, leafy forms of several genera dominate deep regions, and huge boulders several metres across of the genus *Porites* dominate in middle depths. Some genera containing many species such as *Montipora* are relatively inconspicuous but are crucial in reef-building and have members that occupy a wide range of habitats. And, like forests, one dominant form may grade into another. Yet we still talk about 'coral reefs', and this has masked some important changes that are going on today. However, coral reefs of all kinds, like forests, have many unifying characteristics.

Coral biology and growth

Corals are in the animal class Anthozoa, a word derived from the Greek words for 'flower' and 'animal', which describes the appearance of each polyp. Each coral colony consists of numerous small polyps, shown in cross-section in Figure 6. Each polyp has a ring of six (or multiples of six) tentacles surrounding a mouth, leading to the main body cavity. In principle the body is a simple sac contained in a two-layered body wall, meaning they are diploblastic (most complex multicellular animals are triploblastic). These layers are separated by a non-cellular, gelatinous mass of tissue between them called the mesoglea. The different layers contain some elaborate apparatus. In the tentacles, the outer layer, or ectoderm, contains the stinging cells (cnidocytes) that define the entire Phylum, Cnidaria. These are used for catching food.

Millions of symbiotic photosynthesizing algal cells, the zooxanthellae, live in the innermost layer, the gastroderm, sometimes called the endoderm (Figure 7, top). In the polyp's main body, this layer also produces all the digestive and reproductive bodies. The outer layer in tentacles and in tissue that connects separate polyps secretes mucus which is filled with a complex assemblage of micro-organisms that live in association with the coral. These are crucial to the coral animal which, with its associated microbiota, is commonly termed the coral holobiont.

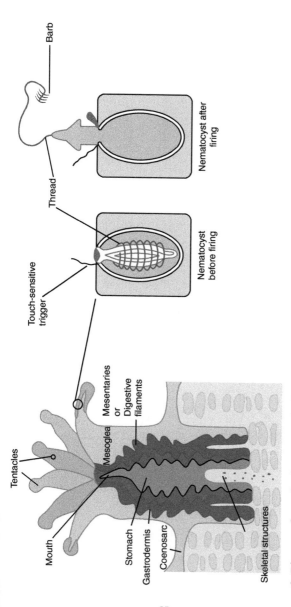

6. Diagram of a coral polyp. A crown of six (or multiples of six) tentacles surrounds its mouth. Its body wall (inset) shows a new and a discharged nematocyst or stinging cell. Nematocysts each have a trigger which, when touched, causes the barb to eject.

Barb

Thread

Nematocyst after firing

Touch-sensitive trigger

Nematocyst before firing

Tentacles

Mouth

Mesoglea

Mesenteries or Digestive filaments

Stomach

Gastrodermis

Coenosarc

Skeletal structures

25

Coral Reefs

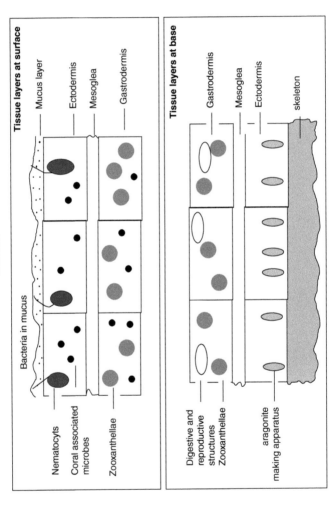

Tissue layers at surface

Mucus layer
Ectodermis
Mesoglea
Gastrodermis

Bacteria in mucus

Nematocyts
Coral associated microbes
Zooxanthellae

Tissue layers at base

Gastrodermis
Mesoglea
Ectodermis
skeleton

Digestive and reproductive structures
Zooxanthellae
aragonite making apparatus

On the underside of the polyp, the ectoderm contains no stinging cells but instead structures needed for depositing limestone in a crystalline form called aragonite (Figure 7, bottom). These structures secrete an acidic organic matrix into the minute space between the polyp and the existing skeleton. This mixture forms a kind of 'scaffold'. Proteins in the matrix bind on to dissolved calcium taken from the seawater. Carbonate ions in the seawater are then attracted to the calcium ions and combine to form aragonite crystals. Sub-micrometre aragonite fibres grow in the organic matrix, and these coalesce and aggregate into well-defined larger units. Each coral species has its own specific pattern of cells that secrete the initial organic matrix, and the pattern of the resulting limestone deposition matches the pattern of cells.

The mesoglea separates these two layers. Although largely non-cellular and mostly water, it contains muscle fibres and nerve bundles and is responsible for maintaining or restoring the shape of the polyp. It also contains 'wandering amoebocytes', which are cells that are important in engulfing (phagocytosing) debris and bacteria, and so are important to the coral's immune system.

A typical polyp thus secretes and occupies a limestone cup, a corallite, which grows upwards and outwards. The fleshy polyps and tentacles generally emerge only at night, retracting into the corallite in daylight, though as with anything biological there are several exceptions.

7. (facing page) Components of a wall of a polyp. Top: tissue wall of tentacles and of tissue connecting polyps. A mucus layer contains huge numbers of microbes—components of the coral holobiont. The gastroderm contains millions of zooxanthellae. Other bacterial components of the coral holobiont are present throughout. Bottom: tissue layers at polyp base. The ectoderm lacks nematocysts but has organelles that secrete proteins that lead to deposition of aragonite crystals.

Some corals are one solitary polyp, but in most species the polyps divide at intervals into two or more daughter polyps that continue to grow upwards until they also divide. There are several variations on this. In many forms, each polyp remains connected to adjacent polyps via thin tissue that also deposits limestone beneath it (Figure 8, top left). Where this is the case, the limestone between the polyps continues to grow upwards along with the polyps. In many cases, when polyps divide they pinch off to form separated polyps with no connecting tissue (Figure 8, bottom right). In others, short groups of polyps develop, which are in one sense a mixture of the two kinds (Figure 8, top right), or polyps remain connected and form long meandering chains. These usually remain attached to adjacent chains (Figure 8, bottom left), earning the name 'brain corals'.

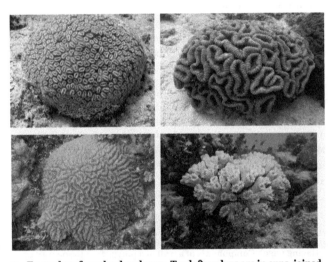

8. **Examples of coral polyp shapes. Top left: polyps are in cups, joined by live tissue (Caribbean *Dichocoenia*). Bottom left: incomplete polyp budding, where series of connected polyps run along each valley (Caribbean *Meandrina*). Top right: chains of polyps again, but each short series is disconnected from neighbouring series (Indo-Pacific *Lobophyllia*). Bottom right: polyps essentially disconnected from all others, and some cups are budding into pairs (Caribbean *Eusmilia*).**

Variations of these lead to a great variety of colony forms: domes, branching thickets, table shapes, columns, and leaves. It matters whether the daughter polyp divides from its parent from within the ring of tentacles or from lower down on the body stalk of the parent: each will lead to different shapes of the ever-growing colony. In species in which tissue continues to connect all the polyps, a touch on one side can lead to a ripple of tentacle retraction across the whole coral, showing that a neural network extends between polyps. In others, such neural connections are lost as the colony grows.

The variety in these structures is used to distinguish between species. The problem with this is that colonies of any one species can vary in shape a great deal according to water and light conditions. It only requires a tiny variation in one aspect—for example, the amount of growth that happens between polyp divisions—to lead, over many repeated occurrences, to strikingly different colony shapes.

The key feature as far as reef-building is concerned is that the coral tissue lays down limestone. Corals thus create their own substrate, and it is this process that ultimately creates reefs.

Coral reproduction and larvae

Reproduction in corals by budding is asexual; in one colony all polyps are genetically identical clones. Asexual reproduction is also used by many species to disperse their colonies, and this is common amongst branching species that fragment to become dispersed by wave action. This may be the main method of colony propagation for some species and, in this way, lagoonal areas may become covered by thousands of square metres of, say, branching staghorn coral. Other variations occur too, an important one in lagoonal areas being that shown by the long polyped *Goniopora stoksi*; this develops 'polyp balls', which are small spheres of skeleton, with perhaps a dozen coral calices containing their fully developed polyps, that develop within the fleshy mass of polyps of the parent. At a certain size, these detach and roll away from the parent colony, a successful means of propagation which effectively expands the reef laterally over sandy areas.

As in all forms of life, sexual reproduction occurs as well. About a quarter of all species are hermaphrodites (with both male and female polyps), but most have separate sexes. Some, such as members of the unattached mushroom corals of the genera *Fungia*, *Ctenactis*, and *Cycloseris* can change sex as they age, mostly male to female, and in the case of *Ctenactis echinata* the corals may later revert back to their initial sex. Most species broadcast both sperm and eggs into the water where fertilization takes place. In some places, and most spectacularly on the Great Barrier Reef, many species do this at the same time on the same night, in a pattern related to season and full moon, producing such vast quantities of sperm and eggs that thick slicks of gametes and fertilized eggs appear on the surface of the water. Sometimes great drifts of this material get washed up onto the beach if there is an unfavourable wind or current direction. It is assumed that this synchronized spawning behaviour has evolved to swamp predators—providing safety in numbers.

Fertilized eggs become planulae and these can drift in the water column for several days or weeks. Planulae are elliptical and can actively swim for limited distances, although they are dependent on currents for their significant dispersal. They respond to chemicals and light in their environment—chemotaxis and phototaxis. This means that they can choose, to a limited extent, where to descend and attach. Chemotaxis might draw them towards locations with favourable chemicals, such as excretory products from their own species, which indicate that the site is probably suited to their kind. Conversely the planula may use chemotaxis to avoid descending and attaching to an area where the chemical composition may be unfavourable, such as on top of a different species where it would be consumed. Phototaxis likewise gives a planula that is ready to settle some ability to determine whether the area of reef over which it finds itself is favourable in terms of amount of light. This requires control of vertical movement to a limited extent, which may permit planulae to remain drifting for a relatively long time while they find a suitable place to settle.

One variant of sexual reproduction employed by some species involves the release of sperm only. The eggs are retained in the parent's body and fertilization occurs internally, with planulae released when they have become more developed. Many of these planulae settle almost immediately, and so tend to grow relatively near their parents.

Coral symbiosis

A remarkable feature of reef corals, and key to their evolutionary success, is the symbiosis that they form with zooxanthellae that grow in the coral tissue itself (see Figure 7). These algae are visible in that their existence in great numbers imparts a greenish brown coloration to the entire reefscape. They are dinoflagellates—microscopic algal cells that are embedded in the tissues of the corals and in many of the shallow soft corals too. In a sense, when you look at a field of coral you are looking at a field of captive, single-celled algae. We do not see great stands of plants on a healthy reef, but photosynthesis is the basis of this ecosystem too.

This symbiosis is likely to be as ancient as the process of reef-building by corals, and is the reason why corals have become such a successful and dominating fauna of tropical oceans. It permits great efficiency in energy transfer from the sun to the reef constructors. The coral benefits from the products of algal photosynthesis. The algal cells use the carbon dioxide respired by the coral tissue to produce oxygen and carbohydrates in their photosynthesis, which in turn are used by the coral, in a captive cycling process that is clearly very efficient. Their densities can reach millions of cells per square centimetre of coral tissue, and this density may fluctuate over time in response to seasonal variables such as radiance and temperature. Although they are dinoflagellates, they lack flagellae when inside the host. Originally thought to be one species, studies have revealed zooxanthellae to be genetically diverse and, as we shall see later, their study has become increasingly important as the environment around corals alters through climate change.

Zooxanthellae are critically important to the process of limestone deposition too. They provide energy to the synthesis of the organic scaffold that is secreted by the coral to make limestone. Deep-water corals in dark environments lack zooxanthellae and so grow more slowly than their photosynthesizing cousins. The end result in sunlit areas is the deposition of massive amounts of limestone—making up the coral reef.

When reef corals reproduce, it is crucial that the juvenile obtains its symbiotic algae as quickly as possible. When a polyp divides asexually, or when fragmentation of brittle forms occurs, zooxanthellae are already there, passed across in the parental tissues. But when reproduction is sexual then the zooxanthellae need to be acquired via the parent's egg or later on from free-living algal cells in the water. Surprisingly perhaps, transmission of zooxanthellae into the egg appears to be difficult, and uptake from the water is more common. The process is poorly understood. Zooxanthellae can aid their own uptake by detecting chemicals secreted by their potential hosts; when they are in the free-living stage they have been shown to swim towards excretory waste products of corals. When they come into contact with the coral they must be absorbed, but not digested. Once established, the stability of the symbiotic arrangement is actively managed by the coral host, which can encourage or slow algal division and growth as it requires, and each polyp can digest or expel excess algae.

This coral–zooxanthellae relationship is therefore key, but it can be affected markedly by environmental stress. Under difficult conditions the coral will expel its algal cells, and because the coral's own tissue is relatively transparent, the white limestone beneath it will then show through, giving a bleached appearance. Recovery of corals may take place if bleaching is not too severe but, following extreme stress, recovery may not take place and the entire coral will soon die. Factors that trigger such stress include extremes of temperature, high or low levels of light, ultraviolet radiation, microbial pollution of various kinds, and pollutants from industrial processes. Bleaching of corals has become extremely significant in relation to elevated sea temperatures.

There are other crucial associations with micro-organisms too. Micro-organisms live on the coral in the layer of mucus that the coral secretes. Most appear to be bacteria and archaea (organisms similar to bacteria but with a differing biochemistry), joined also with fungi, protists, and others, and the whole microbiome is now recognized to be very important. Numbers of these cells dwarf even those of the zooxanthellae: 100 million bacteria can live in each square centimetre of a coral and in its surface mucus. Some are central to the nitrogen cycling processes for the coral, fixing nitrogen, or processing ammonium (a toxic compound of nitrogen) excreted by animal tissues to render it non-toxic. Other microbes process phosphate and sulphur compounds too, the total microbial community performing a nutrient and mineral cycling function, including waste treatment that is essential to the coral. However, as with the algal symbiosis, things can go horribly wrong, such as when excess nitrogen is added to the environment from sewage or agricultural run-off. Some components of the microbiome will multiply rapidly. When this occurs, micro-probes have measured a steep drop in oxygen through the mucus with very little or no oxygen found next to the coral's tissues. The coral tissue will simply suffocate or, if it does not kill the coral host, can make it more susceptible to diseases.

Another abundant microbial component was recently discovered. 'Corallicolids' evolved from photosynthetic cells but later lost the ability to photosynthesize and became parasitic in tissues in the coral's gastric cavity. While abundant, their role is not known. The whole complex of host coral with its array of micro-organisms is termed the coral holobiont.

Coral food—zooplankton

Photosynthesis provides most of a coral's energy needs, but they are carnivores too. The tentacles of most coral species are generally extended only at night. Tentacles are used for feeding, and they are well designed for this. Nematocysts—the stinging darts—are used to capture zooplankton prey (see Figure 6). Nematocysts come in different forms: some are primarily concerned with stinging and injecting toxin, while others act like 'hooks' that latch on to the prey, and then draw the killed prey into the polyp's central mouth. One main benefit of zooplankton capture is provision of additional nutrient compounds that are essential to the needs of the coral. At night, a diver can see how spectacularly rapid the capture of zooplankton may be: if you place a torch beside the coral, zooplankton attracted to the light may touch the tentacle and be immobilized instantly and drawn into the mouth. Nematocysts are fired by contact with their triggers though some corals can sense zooplankton through amino acid residues.

Those corals that do not contain zooxanthellae have the advantage that they grow down to great depths where there is no light, and can grow in cold water too, even up to polar regions. But they grow slowly. The forms are usually minor components of tropical reefs as they are generally outcompeted in sunlit areas, but they are common where it is dark, such as on deep and steep slopes, on the undersides of overhangs, and in dimly lit caves. Many are solitary and small but some species do form large, dense aggregations. Even though these corals depend entirely on zooplankton capture for their nutrition, shallow dwelling forms commonly retract their tentacles during the day like their symbiotic cousins, possibly because of the threat from grazing, and because most plankton only emerge at night anyway. The number of species of azooxanthellate corals is as great as those with symbionts, but they are a generally interesting though minor component of tropical coral reefs.

Coral wars

In normal conditions, there is intense competition for space on the reef between the many species of corals, and between corals and other organisms. While it is easy to see activity, aggression, feeding, and flight amongst fish, for example, corals and soft corals look far more static. But that is only because of our human perception of time. They too compete for space, using several different mechanisms to aid them in their competition for survival. Most of this activity happens at night. You can see its effect by observing apparently bare zones or dead 'halos' around many or even most corals. More spectacularly you may see a distortion or hole in a table coral because lying beneath it, or to one side of it, is a less conspicuous brain coral, for example, that is preventing the table's growth over it (Figure 9, bottom). Soft corals also are commonly seen with narrow bare bands around them, a result of the toxins that they secrete. These animals both prevent other species from overgrowing them and create bare space around them into which they can expand.

This competition for space may take a day, but more commonly two or three weeks, depending on the mechanism that the coral employs. The fastest mechanism involves the dominant species exuding digestive mesenterial filaments onto a neighbour. This is a short-range but very rapid mechanism, which will be triggered within one or two hours if two corals are placed adjacent to each other—in an aquarium for example. It mostly takes place at night under natural conditions, and there is usually a consistently

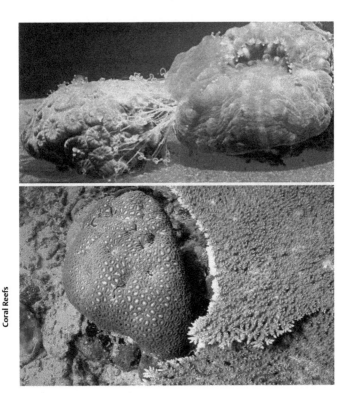

9. Coral competition. Top: a polyp of *Lobophyllia* (right) has extruded its digestive mesenterial filaments over an adjacent faviid colony. By morning, most of the coral on the left will be bare skeleton. Bottom: sweeper tentacles of the coral *Favia* have prevented two different species of *Acropora* table corals from overgrowing it. Normal tentacles of *Favia* are less than 1 cm, but its sweepers, deployed at night, are 10–20 times longer.

dominant and subordinate species in any one pair of species. The dominant coral digests the subordinate species, leaving only bare skeleton by morning (Figure 9, top). A longer-range mechanism involves development and deployment of sweeper tentacles, ten to twenty times longer than normal tentacles, that are heavily loaded with stinging cells. The tentacles sweep over and onto a nearby subordinate coral, stinging and killing it. These sweepers take a couple of weeks to develop following detection of a neighbour, but their long reach is critical. Developing these tentacles obviously requires a considerable investment of energy, so the fact that they go to this trouble shows its importance in maintaining space. A variation on sweeper tentacles is the development of long sweeper polyps, which has more or less the same effect.

Some families of corals may be very aggressive, while others seem to be subordinate to just about every other species. This being the case, in order to survive, the subordinate species must have other attributes such as faster growth or greater reproductive rate.

All this leads to considerable spatial movement on a reef, although during the course of any one dive an observer cannot possibly be aware of it. However, in measurements over two years, I have shown that although the percentage cover of corals, bare space, and so on may remain fairly consistent, beneath any one point there is repeated change from coral to bare space, and back to coral, perhaps of another species, at a rate of perhaps two or even three changes in a two-year period. Movement on the reef may take place in slow motion in human terms, but nonetheless reef surface occupancy is very dynamic.

The soft corals

The soft corals are a widespread group of animals related to the corals and intermixed with them on most areas of reef. These are octocorals, meaning each polyp has eight segments, with eight feathery tentacles—unlike the six or multiples of six tentacles seen in stony corals. As their name suggests, their skeleton is organic and flexible. These corals are especially abundant in the Caribbean (Figure 10). Amongst this group are the sea fans, colonial animals that hold their polyps high up into the water column to feed on

10. An assemblage of many soft corals on a Caribbean reef at about 10 m depth. British Virgin Islands.

passing plankton. Many soft corals also contain symbiotic algae, but most are more dependent on plankton capture. They occupy significant amounts of space on a reef and are ecologically very important. However, they do not help in reef development and their soft colony structures disintegrate after death.

Other calcareous coral relatives

A variety of other calcareous animals may be abundant as well. The organ pipe coral *Tubipora musica* is an anthozoan of a different order, with a hard bright red skeleton made up of small vertical tubes, each housing a grey polyp. The blue coral *Heliopora coerulea* is of yet another anthozoan order, and has a calcareous, bright blue skeleton, though these are covered with brown tissue when alive. This species is most common in shallow water, and both of these species are found only in the Indo-Pacific. Then there are the more distantly related fire corals of the genus *Millepora*, which are in the different cnidarian class Hydrozoa. These are common to both Caribbean and Indo-Pacific reefs. They earn their name because their powerful stinging cells can penetrate thin human skin causing a burning sensation, and although fire coral stings are not usually dangerous, some people may show a severe allergic response to them.

Sponges

On many reefs, sponges are particularly important. Sponges are extremely ancient multicellular animals, whose ancestors extend at least as far back as the Cambrian Period over 500 million years ago. They are enormously varied in shape, even within one species. Some species form huge vases, others are encrusting or consist of solitary pipes. Some ecologically important sponges burrow into limestone and are important bioeroders of the reef. They are important for several reasons. They occupy significant amounts of substrate especially in the Caribbean, and there are over 5,000 species so far identified worldwide. With many more species awaiting identification, their diversity is much greater than that of the corals, which perhaps reflects the longer time over which they have existed. Sponges feed by pumping water through their bodies to extract particulate matter including bacteria down to picoplankton size ($0.2-2 \ \mu m$), and sponges even have the capacity to capture viruses. Their ability to clear water of particles is prodigious: for instance, in the Caribbean a range of sponge species were found to remove two-thirds to nearly all the particulate matter from the water column above them, and sponge communities at 25 to 40 metres deep have been estimated to filter the entire water column above them every day. This capture of plankton is a key link between the water column and the benthic life (life on the seabed). This 'bentho-pelagic coupling' provides an important route through which particulate carbon and nitrogen are channelled between water and the reef.

11. Example of a spur and groove system at the edge of a narrow reef flat in the Pacific. The pink algal ridge is breaking the waves, low tide on a calm day. Spurs extend into deep water down the reef slope.

Calcareous algae and the algal ridge

One crucial group of reef constructors is stony algae. Amongst the red algae are found possibly the most important calcareous species. These stony species thrive on reef crests to create strongly wave resistant ridges (Figure 11). These literally permit the coral reef to persist in the face of storms and strong waves. The algal ridge lies at the seaward edge of the reef flat and marks the point between the reef flat and the slope that plunges steeply. Very few coral species can grow in this most wave-exposed part of the reef.

These calcareous reds actually require strong water movement and high aeration, and some calculations show they are as strong as concrete. From the ridge a series of huge spurs extend to seawards for several metres: this is known as the spur and groove system. The spurs dip gradually into deeper water, and have steep sides, usually with a scoured appearance, and fade out at a depth where average wave turbulence is greatly diminished. The mechanism regulating the spacing of the spurs is intriguing and thought to be some function of the wave energy received at that point, such that the greater the energy from the wave, the larger and more widely spaced the spurs. Whatever their size, the structures greatly reduce the impact of breaking waves in a self-regulating arrangement, and the swash-back of each wave in each groove collides with the next oncoming wave in a way that is both spectacular and yet dissipates much of the wave's energy. This feature is key to the existence of reefs in many oceanic places with high wave energy.

Chapter 4
The resulting structure—a reef

Reef profiles are remarkably consistent in general design, which should not be surprising when it is remembered that they are essentially biological constructions responding to environmental factors.

Reef flats

By far the largest visible expanse of reef when seen from land is the reef flat. Its area usually dwarfs that of the reef slope that descends from it into deeper water and, for a couple of hundred years, the early naturalists focused on the reef flat more or less exclusively, through necessity. Photographs exist of naturalists posing on reef flats in plus-fours, jacket, collar, and tie in the tropical sun, perhaps with a sun-shade and walking stick. This to them was the reef. Figure 12 illustrates cross-sections of two typical reef flats although the extent of the horizontal portion is made narrow to fit the page: reef flats may be over a kilometre wide. Most early naturalists recognized the existence of a profusion of life deeper down, but it remained inaccessible—the *mare incognitum*.

There are reasons for the reef profile. Oceanic tides usually have a range of a metre or more, and at high tide every reef is submerged enough to take a small boat across it or snorkel across it. However,

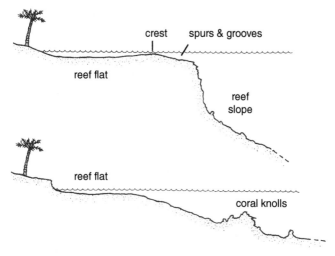

12. Cross-sections of coral reef. Top: exposed or ocean-facing reef, and bottom: a more gradually sloping reef. Profiles can be anything in between, usually a function of overall exposure and geological history of the reef.

at low tide the reef flat usually dries out. This uniform characteristic led to much speculation in the 1800s about why such structures were so consistent.

They are consistent because reef construction can occur only underwater, so structures can grow only up to low sea level. Being so shallow, this area regularly experiences environmental extremes: high temperatures experienced during low tide, severe dilution of salinity by rain during the monsoon at low tide, alternating with dangerously raised salinity under the tropical sun when no rain occurs, not to mention severe desiccation from sun and wind on very low tides as well as damaging ultraviolet light, all ensure that only a small sub-set of reef life has become adapted to live there. Biologically, therefore, reef flats are relatively depauperate. They are an environmentally stressful 'outpost' of the heart of a reef. This factor greatly limited early biologists, because to understand a coral reef from its reef flat alone is akin to trying to examine the Amazon rainforest only from the roadside; you simply cannot see the diversity beyond. Some of the species that have adapted to its extremes may be very abundant, but their diversity is low and many reef flats are relatively devoid of life. It may be an interesting place to research physiology of corals under stress, but this place is not representative of a coral reef.

The types of life that we see on shallow reef flats are either outliers from the main populations deeper down or, in some cases, they are forms that are so well adapted to this location now that they are very rarely found elsewhere, perhaps because they are easily outcompeted everywhere else. Informative examples of outliers from the main populations are the massive shaped colonies of *Porites*, and occasionally genera from the faviid family, which includes the brain and star corals, which can be seen growing on the reef flat in characteristic 'micro-atoll' form. A micro-atoll is a single colony that has grown into a ring with living sides but with a dead and eroding top. They cannot grow upwards any more, only outwards, expanding their girth, and providing clues in connection with the study of sea levels over the life of that coral; fossil micro-atolls have also been found and are used to help in reconstruction of ancient sea levels.

Reef flats therefore develop because coral reefs grow to the low water mark and then can only expand outwards. One well-developed example is seen around the island in the foreground of Figure 3; some reef flats may extend a kilometre from the shore before the reef descends into deeper water. Although they may have a huge area, for corals they are mostly rather marginal habitats.

At the seaward edge of the reef flat, and commonly a little more elevated above sea level than most of the reef flat, lies the reef crest. This is usually marked, and made, by massive ridges of pink, limestone-depositing algae—the coralline algae described and illustrated in Chapter 3. Here there may be extreme turbulence, to the extent that almost nothing else can live there, because the waves break upon it. These structures absorb perhaps 90 per cent of the ocean's energy that otherwise would strike the shoreline of the island.

The reef slope

To seaward of the reef crests, the reef slopes more steeply downwards. Reef corals increasingly grow there until light dims, after which their diversity and abundance declines again. Together with reef flats, this broad structure indicates an actively growing reef. Some reefs, such as atolls, may descend steeply for several hundred metres to far deeper than corals can grow, while reefs on continental shelves may descend gradually to lesser depths. The pattern seen is complicated by changes of sea level over geological time, so the basic pattern commonly will have superimposed upon it marked changes of slope, which complicates the picture.

Environmental conditions on the reef slope are ideal for most kinds of corals and other reef life including soft corals and sponges, and this is where most species may be found. Peak diversity may occur anywhere between 5 to 20 metres deep, depending on the area and overall conditions, particularly overall water clarity. Here, sunlight is still bright enough for photosynthesis but not damagingly intense, and wave energy is not destructively high but water movement is still sufficient to keep sediment from settling. Put simply, most reef species want to live here, where environmental factors are moderate. In order to do so they must show a competitive edge in one or more aspects of their lifecycle—aggressive ability, growth rate, or reproduction perhaps. Coral reef slopes support the richest biodiversity and most varied macro fauna in the seas. It is crowded! Deeper still, at the foot of the reef slope, light begins to dim, and in order to trap light, leafy forms of corals increase and dominate. Wave energy fades to almost nothing, meaning thinner, leafier, and more fragile corals can survive. Their angles are usually such that they trap plenty of light, but their leafy sheets slope steeply enough to shed sediment, and in many parts of the world, the Caribbean notably, the deep reef slopes will be dominated by 'shingles' of corals like the tiles on a sloping roof. Eventually, at depths which may be greater than 80 metres or more, all photosynthetic forms disappear, and the true reef can be said to have terminated, sometimes onto sand, or perhaps onto rocky surfaces (which themselves may be ancient reef), dominated by forms which do not require sunlight.

Cryptic life

A large proportion of reef life is cryptic, that is, occupying crevices or making its own crevices by tunnelling into the soft rock. When it is remembered that about two-thirds or more of the fishes that swarm over a coral reef are carnivorous, many with very wide food preferences, the desire to hide becomes easy to understand. A hidden or cryptic way of life has an obvious drawback, however, namely reliance on food that comes towards it. Hence, many cryptic species are filter feeders, trapping drifting food, and many species prey on those that have done this work for them. Many worms, crustaceans, and molluscs do this, and sponges are exceptionally important filter feeders too. Tiny predators, including crustaceans and molluscs, prowl the maze of subterranean caverns and passages, while other species carry protective shells with them—molluscs and crustaceans being prime examples. For every development, though, counter-measures have evolved: some crustaceans crack or peel open the shells that molluscs create to protect themselves, while some predatory molluscs drill holes in others to get to the flesh beneath. A continual arms race is taking place on the reef, in which one species' defences are countered by another species; and each new kind of attack is countered by new kinds of defence. The 'Red Queen' analogy is very pertinent for reef life, a term derived from the Red Queen in *Alice's Adventures in Wonderland*, who had to run continually just to stay in the same place. All species that exist today have survived as a result of successful strategies in this arms race. This is self-evident in one sense: if the species cannot evolve continuously and keep itself safe from predation and elimination, it becomes extinct—the fate of 99 per cent of all species that have ever lived. And in the most vibrant part of the reef, where environmental factors are relatively benign, competition is intense, and one defence has been to remain as unseen as possible.

Rugosity, a key feature

A consequence of the myriad growth behaviours of corals is a reef which, like a forest, provides a three-dimensional habitat. This is thought to be one reason why reefs support such a high diversity of different forms. Unlike, say, a flat sheet of limestone, irregular coral growth provides considerable volume and countless niches. Most striking in this respect is that most architectural of all coral genera, the *Acropora*. While several coral genera contain some species which develop branches, none match the *Acropora* in their ability to form thickets, miniature forests, and table-shaped colonies, providing much space for hiding, ambushing, or escaping. *Acropora* is the most speciose of coral genera with more than 150 species described. So important is the structure that corals provide that methods have been developed to quantify it and relate the three-dimensional 'rugosity' to various characteristics of reef biology. A diver can see the obvious importance of this by the numerous schools of small fish living in branching coral colonies; they feed on plankton close to the coral but dart back into the shelter of the branches when anything much bigger than themselves approaches. From these small fishes to the largest grouper, many species rest or wait in ambush under tables of *Acropora*, whose wide range of niches provides shelter to both hunter and hunted alike.

The branching structure comes at a price for colonies that display it, the most obvious being loss of physical strength. Because of this, most species of branching corals are found where wave energy is not at its strongest, namely several metres deep, on reef slopes and in lagoonal areas. But there are very stout species of *Acropora* too, in both Caribbean and Indo-Pacific regions. In the Caribbean, the elkhorn coral *Acropora palmata*, which is the largest coral of all in many respects, can dominate the shallowest areas all around islands. Or at least, it used to do so; it has recently suffered devastating diseases that have all but eliminated it—with severe consequences—but it still exists, and is said to be recovering in one or two places. Slightly smaller but equally wave-resistant relatives are found in the Indo-Pacific, including *Isopora palifera*, which similarly occupies shallow and moderately turbulent water, though these too have suffered recently, this time because of warming water.

The massive shaped corals, by which we mean rounded, boulder-shaped forms (even where they are not particularly large) also add three-dimensional structure. Deeper down are leafy corals, extending outwards to capture light. Many have small and widely spaced polyps, relying heavily on light capture for their energy source. Probably all depend to some extent also on organic debris that settles on to their surfaces.

Caves and crevices

Caves and countless hollows in the reef add an additional level of complexity. Whether taking the form of simple small overhangs or deeply penetrating caverns, their fauna is markedly different from that found just outside it. Horizontal runs of caves occur on many steeply sloping reefs, created in the past when sea level was tens of metres below its present level, when that part of the reef was pounded and eroded by waves. Sometimes caves and fissures are solution features caused when sea level was a long way below where it is at present, when copious rainwater etched tunnels that later became drowned when sea level rose again. Whatever the cause, caves show a rich abundance of rather different dominant forms of life, including those usually seen only very much deeper, where they are too deep for most divers to see.

Caves show a marked zonation as well. Around entrances will be the kinds of corals and soft corals that are typical for that depth on the reef, but, further in, as light diminishes rapidly, photosynthetic forms rapidly change to red algae in the case of the plants, from zooxanthellate corals to those without symbiotic algae, and to hydroids and sea fans, and maybe some black corals too—a different group of colonial animals with a skeleton much sought after for polishing into jewellery. The commonest group of attached animals are usually hydroids, most species being feather-like colonies of small cnidarian animals whose powerful stings easily penetrate human skin. Lining the walls of the caves there also may be numerous species of sponges. Amongst and around these sessile animals, and indeed within many of the cavities in sponges themselves, are numerous polychaetes, crustaceans, and other species, specialized for cave dwelling life and different from their more numerous relatives outside.

As this is a dark environment, the fauna attached to the walls of caves is entirely dependent on plankton, both alive and dead, or on the waste products of those that feed on it. The origin of the plankton may be quite obscure. Most presumably are pumped in by water movement from the outside, and so a cave which only has one aperture may be relatively barren. But many caves have heavily fissured walls and roofs, too small to see, but measurable by instrumentation that analyses water flow. Many reefs are remarkably porous, permitting considerable flow of water through them. Most caves experience the passage of thousands of litres of water per day through numerous small channels, pores, and crevices, and to a considerable extent it is this permeability of the reef rock which determines the amount of life that can live in a cave. This can be demonstrated easily, if unscientifically, by a diver inside the cave purging his or her regulator. The air disappears into the roof, and a dive buddy outside the cave might see, after a pause of maybe a few seconds or minutes, streams of air bubbles emerging from many parts of the apparently unbroken reef surface above. Where air can pass, so can water containing plankton.

Caves may not be permanent, and indeed nothing much on the reef is permanent over the longer term. Growth of calcareous organisms and the pumping of sediments into a cave act to close it, even if only very slowly. Alternatively, depending on location and depth, wave action may enlarge it. A cave roof may collapse from the weight of coral above it, and, while not a common sight, it is possible to see places where an avalanche of reef has crashed down-slope, scouring a substantial track behind it which is now filled with sand. I saw an extreme example of this, where the typical reef profile of reef flat, crest, and slope was completely changed by an earthquake 150 years ago, described by Charles Darwin, whose account alerted me to look at that spot. That earthquake took with it not just a large chunk of reef but a portion of an island above it as well, the process being recorded because of the resulting loss of part of a coconut plantation.

Steepening reef slopes commonly overgrow their own foundation, so to speak, and lead to smaller underwater landslides. Reefs have been around for millions of years and these overhangs, caves, and vertical slopes all add different habitats for different groups of species.

Sand creation

A large expanse of many coral reefs is its back reef area, composed mainly of rubble and sand, where few corals grow. This is an integral part of every coral reef, vital ecologically and interconnecting with the reef, in both physical and biological senses. The sand is produced by the reef life and it is truly a part of it.

From one perspective, the principal role of corals is to grow, die, and become broken down into ever-smaller lumps and particles, from which are built up both the coral reef matrix and coral islands themselves. From the moment that they begin to grow, corals are subjected to both biological and physical factors that act to reduce them to sand.

Physical erosion is straightforward. Storms and waves readily break up colonies, especially branching and leafy ones. Fragments produced may initially be fairly large: whole branches, fronds of leafy colonies, and fist- or finger-sized nodules. These, when rolled around by succeeding wave action, fragment further into sand and then silt. Much accumulates behind the reef and in its lagoons, but much also accumulates in the numerous small spaces on the reef itself, where it later becomes consolidated into more durable rock. Rubble that remains on the surface is very inhospitable to most forms of life that require a solid and stable base because its movement in storms has a similar effect to that of liquid sandpaper, producing a surface that is very hostile to most attached forms of life.

The biological attack on coral skeletons that produces sand is more complex and varied, and generally results in creation of much finer particles and silt. Parrotfish swarming in large schools over the reef graze with beak-like mouths, scraping the surface, taking around 1-millimetre-deep bites from the substrate each time. Many boulder corals may be seen covered with these scrape marks, standing out white against the brown background of healthy coral tissue. Parrotfish are responsible for the production of much of the sand on reefs. Their digestive systems are such that they require sand to aid digestion, and it is common to see a stream of fine white sand being defecated from a parrotfish as it swims.

Far less visible is bioerosion by cryptic forms of life, usually very small animals, and by some plants as well. Although largely unseen, the cryptic reef fauna are estimated to have a biomass as great as that of the more visible fauna on the surface, and have a much greater diversity of species. Small, bioeroding organisms tunnel, bore, and chisel into the relatively soft limestone rock. Many bivalve molluscs use their shells to scrape out a tunnel in which they live. Other organisms secrete mild acid to dissolve the limestone. A particularly important bioeroding group is the sponges. These have no moving parts and so rely entirely on dissolving the rock. Some develop extensive but thin sheets of tissue covering the surface, distinguished by many irregularly spaced, small volcano-like structures, which are the exhalant pores of the sponge, whose main biomass lies beneath in an increasingly hollowed out coral. Some large sponges of the genus *Cliona* (Figure 13) can erode many kilograms of substrate in a year from corals that they have colonized. Net bioerosion from sponges has become a dominant factor in many reefs where corals have been killed for whatever reason, but most corals on a healthy reef are not colonized in this way.

13. A coral colony being destroyed by the sponge *Cliona*. The sponge, bright red, is encrusting only on the surface; most of the sponge is inside the increasingly hollowed-out coral limestone. Caribbean example.

Plants also bioerode. Filamentous green algae, especially, work their way in, often using the minute, naturally existing pores in any coral colony. Being photosynthetic, they need to remain in the top few millimetres where they still receive some illumination. The translucent quality of corals' aragonite limestone means that light can penetrate sufficiently through both a thin layer of coral tissue and a couple of millimetres of limestone for sufficient photosynthesis to continue. A section through many massive corals will show a thin green band that is indicative of the presence of such algae. One of the main groups is a genus called *Ostreobium* and, because this group lives so near the skeleton's surface, it has been suggested that they may even contribute in some way to the nutrition and gas exchange of the coral, and some view this as part of the coral holobiont.

Overall rates of erosion of reef and corals may be close to the highest levels of calcification or growth that have been recorded. Accumulation of limestone may typically be around 6 kg per square metre, every year, and erosion a little less on a healthy reef, so there is net growth, if only just, which is why coral reefs tend to reach the surface in the first place. But any interruptions to growth, or increase in erosion, can easily result in an imbalance, leading to destruction of the reef. Another factor is important. Polyps cover the surface of a living coral, and polyps are carnivorous. Any unfortunate larva of a tunnelling or boring organism trying to settle on a live coral is consumed, so that the live surface of any coral is a most effective barrier to penetration of its skeleton. Mortality of the coral tissue exposes the skeleton to attack, and leads to greater sand and sediment production. It is not only a psychological gloom that makes a diver think that a dead reef has poor water visibility, but also the increased quantity of very fine silt suspended in it that has originated from such erosion.

Many other groups of organisms produce sand, some in significant quantities. Foraminifera, or forams, a group of protists, have shells usually of limestone that form sand grains less than a millimetre across to perhaps a centimetre, depending on the species. Benthic species of this group may be so abundant that they comprise most of the sand over extensive areas. Colonial species of forams create encrusting sheets and play an important role in helping bind together the sand and rubble beneath them, and together forams can create huge sand banks.

Other groups of animals contribute to the sand in relatively minor ways and may be distinguishable by having colour, sometimes vivid, in the otherwise pure white sand grains. Mollusc shells, echinoderm tests and spines, and skeletons of the blue coral *Heliopora* and the red organ pipe coral *Tubipora* all become part of the sand, adding flecks of reds, browns, and blue.

The sand produced plays constructive roles on the reef, an aspect commonly overlooked but which is vital to the growth of the reef. Sand and silt settle into crevices and pores, sometimes forced in deeply by wave action. These particles fuse and slowly turn the reefs into consolidated limestone rock that is more durable than the coral colonies from which the particles were derived. A cross-section of an exposed fossil reef will reveal as much or more of this amorphous looking solid limestone matrix than recognizable coral colonies. How these particles consolidate or become fused into solid rock has something to do with microbially mediated changes in pH on the surface of the grains, but the total mechanism is largely unknown. The result, which can be seen in an exposed fossil reef, is a structure of numerous dead coral colonies of many species surrounded by and embedded in a durable rock matrix. It is this, principally, that creates the geological structure of a reef.

Reef sand is also made directly by several kinds of plants from all three major categories of large algae—the greens, reds, and browns. The most important of these is the green *Halimeda*, an unmistakable plant consisting of chains of small limestone discs covered with living, green tissue, each disc being round or kidney shaped, and a few millimetres to a centimetre wide or more depending on the species. Each disc is mostly limestone, and each frond may produce a new disc every day under ideal conditions, with one plant developing dozens of fronds. Connections between discs are fragile and break easily; the tissue dies and the discs become a readily recognized part of the sand. Their production of sand is prodigious, such that many Caribbean beaches are largely made up from this and, in the Pacific too, enormous mounds have been discovered in depths of several tens of metres. Vertical underwater surfaces may be covered with *Halimeda* plants from which stream billions of discs daily, destined to become a dominant part of the sand and an important part of the whole reef complex.

A few brown algae also secrete limestone within their fronds though nowhere near to the extent of *Halimeda*. The genus *Padina* is typical of these, though the quantity of limestone it produces is much smaller and very fine. Red algae likewise include several genera with similar limestone secreting properties, though the most important algae in this taxonomic group are those stony genera that develop the algal ridges and spurs described earlier.

Sand life

The quantity of sand produced on a reef generally exceeds several kilograms per metre square per year. While much becomes entrained in the reef, much more drifts into back-reef areas to create expanses larger than the reef itself. Important sand-sorting processes happen continuously, driven by waves or currents whereby sand is continually graded from coarse to fine. Found immediately behind the reef is generally the rubble and coarse sand, leading to ever finer sand and finally possibly to calcareous muds. This sorting occurs primarily because finer particles remain suspended in water for longer and settle out more slowly, so are carried further in the water flow which is driven by waves breaking on the seaward edge of the reef. It is commonplace for reefs to grow into prevailing wave energy, so that there is almost continuous movement of water from the reef front or reef slope, across the reef flat, to areas behind. The sand may become piled up to form a beach around an island, sustaining that island or even creating a new one. Currents across the top of a reef may commonly flow at more than one metre per second, more than enough to carry suspended sediments considerable distances, but when deeper water is reached the velocity immediately slows, allowing grains to progressively settle out according to size.

This sand is far from lifeless, and different sized sand particles support different assemblages of species. The first point of note is that the sand underwater very often appears brown, grey, even purple, and mottled—not white as when it is piled up to form a beach. The colour is caused by expanses of filamentous algae and microscopic forms such as diatoms growing over the surface, photosynthesizing, and providing food for numerous other organisms. Many of these also fix nitrogen. Commonly mixed with these are enormously productive patches of larger algae and seagrasses. All of these add organic matter to the substrate, which is consumed by numerous larger forms of life. Many species contribute to the sand itself. The most numerous animals are small, for instance the so-called micro-molluscs, which are the size of sand grains themselves. Much more conspicuous users of sand may be sea cucumbers of the echinoderm family Holothuriidae, which feed on the organic matter by ingesting vast quantities of sand, digesting its organic content and excreting the rest. In those increasingly rare places where they are not collected for Asian food products, their numbers might exceed two or three in every square metre; they are a crucial component for 'cleaning' the sand. Where they are heavily collected, organic matter can build up to levels which lead to anoxia.

Numerous molluscs, including many predatory species, burrow within the sand, some surfacing only at night with their siphons extended above the surface during the day. Many flattened forms of echinoderms called sand dollars and heart urchins also burrow. Some of the most conspicuous forms are worms, notably polychaetes, many in tubes with one end exposed to the surface. Several fish burrow into sand. Striking examples are the garden eels, shy creatures forming a sub-family within the conger eel Congridae family, whose colonies can cover thousands of square metres, and which withdraw tail first into their burrows at the slightest disturbance, including the approach of a diver. Burrowing shrimp may be common and conspicuous sand dwellers, as are rays, partially buried beneath the top layer, with just a pair of eyes visible.

Burrowing by such species is enormously important because it ensures that sub-surface sand and silt is regularly mixed with surface sand, bringing it up to the top where it is exposed to oxygenated water in a process that prevents an anoxic layer developing below the surface. Experiments in which black sand is spread over areas of white coral sand have shown that a thorough mixing of surface with sub-surface material takes place within just a few days.

Most families seen on a reef have relatives adapted to living on and in the adjacent sand. This area remains much less studied than the reef itself, which is unfortunate as it is a very important component of the wider reef ecosystem. The sand may build up an island, something of huge commercial and tourist importance, not to mention in providing a home for people, and huge sandy expanses provide the principal feeding area for several species of fish used by local communities. Such areas should be viewed as being an integral part of the reef. Many fish of commercial importance migrate back and forth between sand and reef, some on a diurnal basis and others on a seasonal basis, to forage for food, forming an ecological link between the reef and its adjacent sandy areas.

Chapter 5
Microbial engines of the coral reef

We know that the symbiotic algae in corals, soft corals, and their relatives are a crucial source of fuel for the reef, and that there is a tight recycling of nutrients between the coral animal and its symbiotic algae. There is little 'leakage' out of the coral–zooxanthellae symbiosis. But several other important engines of coral reefs exist, and are the largely unseen mechanisms by which organic material and energy produced by photosynthesis is transferred along various food chains, from plants to the top carnivores. Transmission of the organic matter, and of the energy, through the system depends to a large extent on the microbes of the reef. Microbial life is key to both the overall high productivity and to the diversity of the reef.

Microbial life, including plankton, can be divided into several groups for ease of explanation. According to size, starting with the smallest, these would be viruses; bacteria and archaea which are prokaryotic; protists; and plankton with more advanced, eukaryotic cells. The latter includes phytoplankton and the larger zooplankton, including larvae from larger animals.

Viruses, the smallest of all, cannot live independently. To reproduce and survive they must invade a much larger host cell, and hijack that cell's genetic apparatus to divide and reproduce. They are tiny, ranging from about 20 to 250 nanometres. Marine viruses are still not well understood and remain a rather enigmatic component on reefs. They occur in densities of up to 10^8 in each millilitre of seawater, and may be more abundant still by a couple of orders of magnitude on the seabed. They probably infect all organisms. While the filter feeding apparatus in some reef sponges can extract some viruses from the water directly, viruses mostly have their greatest effect on many ecological processes by regulating populations and communities, just as they do on land.

Bacteria and archaea are larger, though still much smaller than the cells of 'higher' and more complex multicellular organisms. They are prokaryotes, meaning they have no nucleus. They are 0.2–2.0 micrometres (μm) in size, and when suspended in water they form part of the so-called picoplankton. Together bacteria and archaea are the most abundant part of the plankton. Archaea and bacteria are of similar size, causing archaea to be considered initially as part of the bacteria, but archaea are now placed in a separate Domain; indeed, the archaea have some very different genes and metabolic pathways that suggest they have closer affinities to cellular and higher organisms than they do to true bacteria.

Bacteria form common symbiotic associations in the mucous layers of corals, and many more live inside coral tissues (shown in Figure 7), being a key component of the coral holobiont. Some of these bacteria have been cultured, but a better understanding of their diversity and abundance has been revealed by sequencing their genetic material. There are certainly several hundred different bacterial species, and of archaea, in the coral holobiont.

In terms of diversity, turnover, and metabolic activity they exceed all other components of a reef, and they break down organic matter and recycle it, mineralizing some nutrients, growing on the material in clumps, then becoming food themselves. Their doubling time can be prodigious: ten doublings per day have been recorded. They can show very rapid daily fluctuations in density as they respond to new food sources, being most numerous during daylight when they respond rapidly to photosynthetic products of algae and mucus from corals. Benthic bacteria are equally abundant and diverse, with a productivity on a reef floor of half a gram of carbon per metre square per day.

These microbial communities, especially bacteria, are key to the recycling of carbon and organic compounds on reefs, and this is known as the 'microbial loop': the invisible aspect to how a reef functions. Key nutrients such as phosphorus in the water over a reef can be turned over in minutes, and even complex organic materials are recycled in hours or days. For many years it was considered paradoxical that coral reefs could exist as oases of high productivity in the nutrient-poor vastness of the open ocean, but it is clear now that one reason why this is possible can be attributed to the microbial loop.

Direct consumption of microbes that have grown on waste, dead, and dissolved organic matter is fundamentally important. Bacterial clumps around such waste become the food source for many filter feeders and for millions of tiny scavengers on the seabed. Many of these scavengers are single-celled animals, but others are minute, multicellular forms.

The cyanobacteria, formerly called 'blue-green algae', are a major part of the 'microbenthos'. Over half of the reef surface may support these agents of primary production, and they may occur in densities of millions per litre of reef sediment, commonly giving a dark colour to the otherwise white coral sand and providing substantial amounts of biomass in beds of seagrass and large algae too. They are sufficient to contribute 20–30 per cent of the total primary production of coral reefs.

Protists are larger than the bacteria and archaea, and are single-celled organisms. Although not animals or plants they include plant-like forms found in the phytoplankton. They are eukaryotes, meaning the cells have nuclei and organelles, as found in all multicellular complex life. When in planktonic form, they are nanoplankton (2–20 μm across) and microplankton (20–200 μm across). These can also be very abundant, and all assist in scavenging organic material. They are relatively scarce in very clear water but occur in high densities in the water column in reefs nearer land, where perhaps there is nutrient run-off acting as fertilizer. Diatoms and dinoflagellates are important components. Some dinoflagellates are especially known for their symbiosis with corals, but many are entirely planktonic and many more live on surfaces of rock, seaweeds, and seagrasses. Some dinoflagellates are motile and entirely predatory. They have lost their chloroplasts, and feed on other protists, using their two flagellae to swim. Sometimes free-living forms undergo blooms, forming 'red tides', when they may produce neurotoxins that result in massive fish kills and produce ciguatera poisoning in humans who eat fish contaminated by them.

Several hundred species of dinoflagellates and diatoms can live in the top centimetre or two of a sandy patch behind a reef, living at densities of two to three million diatom cells in every millilitre of sediment. This provides up to a gram of chlorophyll in every cubic metre, and they have a significant total biomass. These microscopic algae may contribute a quarter to a third of the total primary production on a coral reef.

Single-celled animals, like protists or proto-zooplankton, are the next step up the microscopic food chain. Several are major grazers of phytoplankton and benthic forms of diatoms and dinoflagellates. Grazing of phytoplankton by vast numbers of protists is a major link in the coral reef's trophic web, responsible for much of the carbon transfer to larger organisms. They are diverse, occurring in both seawater and sediments. Ciliates are important too, and many of the larger forms also contain symbiotic dinoflagellates, so as well as consuming smaller plankton, some also increase primary productivity.

We have already come across one important group of protists, the forams. These are larger than the above groups, at 1–15 mm across, and there are both planktonic and benthic forms. These amoeboid organisms exist in huge numbers on the seabed. Many develop very elaborate external skeletons which, in some reefal areas, also produce significant amounts of sand, and their skeletons have been very useful in tracking past climatic changes. Many contain symbiotic algae too. Via these forms too, much of the energy and carbon moves up the numerous food chains of coral reefs.

Zooplankton

The major mode of feeding by a large proportion of reef animals is filter feeding—extracting particles from the water. Corals do this using their tentacles with stinging cells, even though corals may obtain most of their energy from their photosynthetic symbionts. Soft corals also do this, but they need a greater proportion of plankton to supply their energy needs. Sponges, and many bivalve molluscs, feather stars, countless worms, and basket stars, amongst many others, feed heavily on the plankton, both the kinds described above and especially on zooplankton, either by pumping water through their bodies and sieving out useful material or by draping tentacles in the water currents to trap the particles as they pass. Some species are able to discriminate not only the size of plankton but its material too—a particle of fine sand has no nutrient value and needs to be rejected, whereas a zooplankter of the same size is retained. Some of the smallest fish are 'plankton pickers'; these literally see and pick individual zooplankton suspended in the water, while larger sea creatures up to and including manta rays and whale sharks simply scoop in huge volumes of water and filter it. Large zooplankton feed on smaller zooplankton, and smaller zooplankton feed on phytoplankton.

Permanent zooplankton. Except during episodes of mass spawning by corals and their relatives, the most numerous components of zooplankton on most reefs are permanent, some living in the water column and others living between grains of sand. This permanent zooplankton includes a wide range of crustaceans, polychaetes, and nematode worms. A large group are copepods—tiny crustaceans—whose total biomass may exceed that of their larger, familiar relatives such as crabs. Most of this minute fauna can grow rapidly in response to the bacterial food source, while large scavengers such as sea cucumbers feed on this material as they ingest large quantities of sand.

Many amongst the permanent zooplankton are demersal, living in crevices in sand and rubble during the day, and rising from the reef into the water column at night. The numbers of demersal zooplankton usually greatly exceed those of zooplankton that are entirely pelagic, and as a consequence demersal zooplankton are a major source of food for plankton-feeding species. Measurements of their density in sandy substrates show extraordinary numbers. Red Sea sand contains many thousands of animals in each square metre, of which three-quarters or more migrate upwards through the water column at night. Usually they do this within an hour of sunset and return to their crevices in the sand and rubble near dawn. It is this group of plankton that is the target for many benthic-dwelling filter feeding organisms: corals extend tentacles at night and feather stars and basket stars unfurl their arms to catch them: night time is their time to feed. Several major groups, again in the Red Sea, have been shown to migrate vertical distances of 25 metres from their rubble substrates to near the surface of the water, creating a steep gradient in the concentration of zooplankton over a reef. Remarkably, demersal zooplankton from many taxonomic groups show movement patterns that are extraordinarily well synchronized, emerging from the reef within a few minutes of sunset all year round, and returning to the benthos about an hour and a half before dawn. It is thought that just after sunset planktivorous fish are much less efficient at capturing zooplankton and most corals have not yet expanded their tentacles, and these factors offer a brief window of time in which the zooplankton can escape predation in that most dangerous layer just above the substrate. Presumably, predator avoidance is the main factor determining the precision of this timing. In total, this demersal plankton is a significant component of the trophic web on a reef. Although the animals are largely unseen to the naked eye, their effects are abundantly clear in that they support much of the visible life.

Temporary zooplankton—larvae. Larvae of larger animals form part of the plankton too. The purpose of the larval stage of a sessile animal is dispersal. The largest larvae are usually those of sessile species including corals. Many can swim to a certain degree and can control their vertical movement in the water column, regulating to some extent their own dispersal. The length of time that larval stages remain in the water varies considerably, though two to four weeks is common amongst fish and corals. Very mobile larvae such as those of the surgeonfish can swim continuously for nearly 200 hours, covering distances of tens of kilometres. During that time these zooplankton may move several hundred kilometres in water currents, but there is also evidence that some species can extend this duration considerably, for example if they do not detect suitable substrate on which to settle or if food is very scarce, in which case they can enter a dormant phase until conditions improve. This phase may last for many months—some have even been referred to as being 'virtually immortal'. This has important consequences for the dispersal of reef species.

A factor important to the survival of planktonic larvae is their sensory organs. Many respond to gravity (known as geotaxis); those that are strongly geotactic will settle quickly, with the advantage that they settle near their parents and thus in a habitat that is likely to be suitable for them, but with the disadvantage that dispersal is limited. Others respond by swimming upwards, which has the converse benefit of greater distance dispersal, but increases the risk of moving into an unsuitable habitat—and, of course, prolongs the risk of being eaten by a plankton feeder.

Many reef-dwelling species, including coral larvae, settle preferentially onto limestone and onto calcareous red algae in particular, using chemosensory mechanisms. Another intriguing attractant to some larvae is sound, notably the sound of the reef, which includes components of breaking waves, and the clicking and snapping noises made by reef animals; a reef-dwelling species would be drawn by these characteristic sounds to reef habitat as opposed to, for example, expanses of sand around the reef.

All these groups together form the primary fixers of carbon, cyclers of nutrients, and conveyors of organic material through the web of life, from initial microbial decay of debris to the larger and more visible life forms.

Other microbial symbiotic connections

Other than the key coral–zooxanthellae symbiosis, coral reefs show very many more symbioses between quite different forms of life which are also key components of material and energy transfer. Sponges provide a good example, and are viewed by some scientists as being colonies of symbionts. The sponge animal forms the framework, an ancient form of multicellular life but one that probably could not exist at all without symbionts. Most if not all shallow reef sponges contain huge numbers of symbiotic or commensal algae and bacteria, especially cyanobacteria, harbouring these both within and between their cells. Multicellular algae, particularly the reds and greens, also form remarkable associations with sponges. But all the most important symbionts are, again, micro-organisms. Many fix nitrogen from the water, with important consequences for the availability of this critical nutrient. When it is considered that sponges at some depths and in some areas such as the Caribbean may occupy as much substrate as do the corals, it is clear that they are as important to the ecology of the reef as are corals themselves. The symbionts increase the ability of sponges to grow, to capture particles from the water column, to stabilize loose substrate, and, depending on the species, to bore into the reef and cause bioerosion. With sponges as much as with corals, symbiosis is an essential mechanism that tightly cycles nutrients and metabolic gases as well as producing higher photosynthetic products. Some sponges also benefit from certain cyanobacteria that are highly toxic. These defend sponges from predators—indeed some Caribbean sponges are toxic even to human skin.

Bacteria form common symbiotic associations in the mucous layers of corals and many more live inside coral tissues. Some of these bacteria have been cultured, but a better understanding of their diversity and abundance has been revealed with the advent of DNA examination techniques. Analysis of just a handful of coral samples has shown several hundred different bacterial species, about half of which appear to be new species, and it is thought that several thousand different kinds exist. Their diversity in any one area may parallel the diversity of corals themselves, in that high latitude reefs have been shown to have fewer species than those closer to the equator. Probably as common and as diverse are members of the archaea group, micro-organisms that are quite separate from the bacteria, reportedly existing in densities of tens of millions on every square centimetre of coral surface. Little is known about them or how they affect the food chains that power the reef. However, their huge numbers and ubiquitous nature suggests that they will be shown to be important.

Productivity from seaweeds. On a healthy coral reef there are relatively few visible, large 'plants' such as are common on land and in colder seas, and most of the energy pathways that drive the reef system have involved or been based on microbial organisms. The major exception to this comes from small and filamentous algae, mainly green forms growing between the corals and soft corals, and these are a major food source for grazing fishes. These seaweeds grow rapidly in all areas, but there is usually not very much of them to be seen on a healthy reef because they are grazed as rapidly as they grow by the numerous grazing fish, sea urchins, and others. Their standing crop may be small but their turnover is rapid, as can be seen if an area is experimentally caged to exclude grazers. In such sites, the substrate soon becomes covered with rich seaweed growth, mainly filamentous forms. This demonstrates significant transmission of material and energy along this food chain too. These beds of 'turf algae' are also a prime habitat for many other forms of life.

The rate of turnover of these microscopic forms of life is extremely rapid. It is not too much of an oversimplification to suggest that they power the reef. Referring to the coral veneer as well as microbial components, biologist Bruce Hatcher described it thus:

> Coral reefs are gigantic structures of limestone with a thin veneer of living organic material—but what a veneer! Everything that is useful about reefs (to humans and to the rest of nature) is produced by this organic film, which is approximately equivalent (in terms of biomass or carbon) to a large jar of peanut butter (or vegemite) spread over each square meter of reef.

The quantity of living organic tissue on the reef at any one time is not necessarily very great but its rate of turnover is immense.

Chapter 6
Reef fish and other major predators

The kaleidoscope in the water

Perhaps the most immediate visual impression for a snorkeler or diver visiting a healthy reef is the immense number of fish in a variety of shapes, sizes, and colours. Some, especially the smaller ones, are very timid, never venturing far from the protective branching corals, while others, such as parrotfish, swarm over reefs in great schools scraping algae and other food material from both dead rock and living coral. On an unfished reef, thousands of fish hover in the water above the reef. No other natural habitat in the ocean shows this diversity and abundance of fish. Clearly, there are strong driving factors leading to the proliferation of these numerous species of differing size and body shape. Many fish provide food for coastal communities too.

Many smaller species are coral obligates, meaning they can live only in, or feed only on, particular kinds of coral, especially branching forms. This naturally constrains their overall abundance and distribution. When the corals on which they are dependent die, obligate fish species are the first to show population collapse. The association that any particular fish species has with a reef may be more or less permanent, or it may occur for only part of the lifecycle of the fish, and it might be food or habitat related.

As with corals, there are major geographical influences on the diversity patterns of reef fishes: there are quite different assemblages in the Caribbean for a start, and there are many more species in Southeast Asia than in the Eastern Pacific. Relatively isolated areas such as the Red Sea have many endemic species not found anywhere outside that region, while some species range across entire oceans. There are patterns of diversity correlated with latitude too. Furthermore, on any one reef the diversity of the fish assemblage increases with the overall complexity or rugosity of that reef. Generally, the more complex the topography of habitats, the more complex and varied are the assemblages of fishes that utilize them. And again, as with corals, the various species have adapted to different depth ranges, with some thriving in the shallow surf zone, many requiring vast expanses of sand over which to forage, and some being restricted to deeper areas. As with the corals, most species prefer intermediate depths, while large predators hunt across all depths.

Feeding ecology of reef fishes

The hundreds or thousands of fish species found in a reef complex show most possible feeding types, very often with one type grading into another. Our human mind prefers to categorize particular species as herbivore or carnivore, for example, but many species cross these classic boundaries. Some fish families contain species with a more or less consistent feeding habit (e.g. most shark species) but other huge families, such as the parrotfish and the wrasse, contain species with everything from plankton feeders to detritivores, herbivores, mollusc feeders, and others. The same species may have a different diet in different parts of the world. Fish size itself is an unreliable indicator of trophic status. Some carnivores are just a few centimetres long while the largest fish of all, the whale shark, feeds mainly on zooplankton.

Detritus feeders are very abundant; detritus is concentrated on the seabed, mostly amongst the filamentous or turf algae, and this is where an enormous part of detrital processing takes place. Because it is so organically rich, it is a primary grazing site for detritus-feeding fish. Detritus in algal turf may comprise up to about 80 per cent of the organic matter present, which provides more nutritional value than the algae itself. From forams to fish faeces, algal turf is organically rich. Fish from several families feed on this mix and in the Great Barrier Reef these constitute about 20 per cent of the different species and nearly half of the biomass of fish in that habitat. The fish that feed here, such as parrotfish, are extremely conspicuous, sometimes foraging in large schools over the seabed.

Plankton-feeding fish are common also. Amongst these are the smallest known: 'plankton pickers' which hover just centimetres above their bolt-hole and grab individual, large zooplankton. Because much zooplankton commonly emerges at dusk and returns to crevices during the day, many of the fish that feed on them are nocturnal, noted for their enormous eyes. One of the smallest fish known, *Trimmaton nanus* (sexually mature at only 8 millimetres long), is a plankton picker.

Herbivorous fish diversity is fairly high, making up perhaps a quarter of the total species present. They provide considerable biomass. They are mostly found in shallow, well-illuminated water, since light is required by their plant food. The raspers and grazers such as parrotfish with their strong beaks take in prodigious amounts of limestone along with their organic food; this scraped limestone might be up to three-quarters of the total mass ingested and is important to their digestion. In fact, their defecation of it, after they have extracted the nutritious component, makes up much of the sand on the reef—up to one tonne of this material can be produced in a year by a single, very large parrotfish, and a school of smaller parrotfish produces more than this, all from rasping at the reef. The bioerosion of the reef that this causes is therefore considerable, and these rasping fish have a major structural function on a coral reef. The scraped areas are quickly colonized by benthic life, including, crucially, coral larvae, which will go on to form a new generation of corals.

Since the algal turf contains so many micro-organisms, fishes feeding on them are certainly not pure herbivores. Some fishes favour certain groups of algae while others are more generalist. Some, such as several damselfish in the large Pomacentridae family, become very territorial over favoured patches, vigorously repelling other encroaching fish as they try to keep a few square metres for themselves. Their behaviour is very much linked with their diet. Once, when I staked out quadrats for coral measurements using steel pegs, damselfish promptly made several stakes the centre of their own guarded patches so that, on returning three months later to measure coral cover and recruitment, I found damselfish were 'farming' these experimental areas, which were now smothered with algae. Indeed they even tried to repel me—these little fish are instinctively fearless, even though they are easily overwhelmed by, for example, a school of parrotfish swarming through the area. Some herbivorous fish, like the detritivores, forage over large areas, some showing migration patterns between reef and adjacent sandy areas, perhaps on a diurnal basis or perhaps with cycles linked to reproduction. Although turnover of algae may be very high, at any one time the standing stock, or biomass, is generally low because of continued cropping.

Fish that feed on invertebrates include generalists that will eat most things they can find. Other fish are more specialist. Feeding on invertebrates that live just under the surface of the sand requires very different skills and feeding equipment from chomping on dead coral rock to obtain the invertebrates living within, and leads to very different adaptations. Goatfish, comprising the Mullidae family, feed on sand using a pair of sensitive barbels ('feelers') beneath the chin with which they can sense their prey. Many have a specially shaped mouth with which to get at their prey in rubble. Rays have mouths under their heads and use their wings to disturb the sand to shake up their prey. Some fish feed on sponges, which also requires specialized digestive processes to cope with both the sponge's toxic chemicals and the sharp siliceous or calcareous texture of the sponge matrix.

Some fish feed on corals (corallivores). The very attractive butterflyfish are a classic example, and these make up about half of the estimated 130 species of corallivorous fish known to exist. Some butterflyfish exist exclusively on just one or two species of corals and have developed elongated mouth shapes to facilitate picking off individual polyps. About one-third of corallivorous fish feed only on corals, obtaining nutrients from polyps and coral mucus, while the remainder also feed on other invertebrates when these are encountered.

Carnivorous fish at the top of the food web may be the most numerous in some areas of reef. Many detritivores are at least partly carnivorous too, ingesting small worms, crustaceans, and other small animals as they forage. Some unexpectedly docile but large sharks, such as the nurse shark, feed on crustaceans and molluscs. But most piscivores, the fish hunters, sit at the very top. These are generally the fastest swimmers as they have to catch their fast moving prey. It is thought that the common tendency of schooling by many potential prey species of fish is either because schooling helps to confuse the hunter or because it provides a form of safety to individuals—much as with herds of wildebeest in Africa. However, some predators school too: some barracuda and jacks hunt in coordinated packs, and a classic sight on a reef is schools of very fast swimming jacks, barracuda, and others 'carving up' schools of prey species, isolating some individuals before snapping them up. The modes of operation of piscivorous fish, and their range of body shapes, are as varied as their prey, and include stalkers and ambushers, as well as the active, sleek, and fast hunters. Some like the grouper simply lie in wait and gulp passing prey by opening their mouth rapidly, causing an inrush of water and, with it, the hapless target (Figure 14).

14. A large grouper, camouflaged, and an ambush predator, waits motionless for smaller fish to pass very close to its large mouth. They are a common sight in caves and under overhanging corals.

The piscivorous sharks are amongst the most visually impressive hunters on the reef. They have astonishingly acute olfactory senses and organelles that detect vibrations and electrical signals in water, including those coming from an injured fish. Sharks have an important 'top-down' controlling influence on reef ecology, as do many other large piscivorous groups, but their place in our folklore is unparalleled because of the very occasional attack by some species on people; usually these attacks have been on swimmers rather than on divers, except in cases where the latter have been spearfishing and are hanging on to injured and dying fish.

However, sharks are currently experiencing appalling destruction at the hands of humans in all oceans of the world, sometimes for consumption of their flesh but often, more wastefully, just for their fins. Consumption of shark-fin soup is enormously popular in Asia. One hundred million of these high-level predators are taken every year, with enormous ecological consequences. Unfortunately, the unfounded perception of the danger sharks pose to humans is not really changing. Ignorance about sharks and their important role is remarkable: even at a marine science conference in China, I was served shark-fin soup (and my refusal, as a guest of honour, caused bafflement as much as consternation). About a hundred shark attacks on people are recorded each year (one millionth of the number of sharks being killed) and about a fifth of these are fatal. Most of these attacks occur in cool water areas and estuaries—not in reef waters.

Piscivores can feed on other carnivorous fish as well as on herbivorous fish, so they cannot be conveniently placed on one simple level of the food web. In fact there is no single classical structure of proportions of fish on a reef. Taking reefs that have highest fish biomass, in Kingman reef in the Pacific, the apex predators are mainly sharks; in Chagos in the Indian Ocean, on the other hand, the large apex predators include far greater numbers of other groups such as snappers, jacks, and groupers, with far fewer sharks. It is not known to what degree this is the result of illegal fishing and poaching, which undoubtedly occurs in both places, all such fish being highly desired by humans for food. And where there are human settlements nearby, there is invariably the greatest reduction in all top-level carnivores, which swiftly causes ecosystem distortion. The removal of these carnivores might be expected to lead to 'predator-release', in other words, an increase in numbers of their prey, such as herbivores. However, the herbivores are also fished in large numbers.

While a healthy coral reef does have, like any ecosystem, a roughly pyramid-shaped trophic structure, the fish component alone does not necessarily have this structure. Indeed, the 'fish pyramid' on its own can even be upside down. However, one important point about a coral reef is that, as with plant biomass, it is not only biomass per se that is important but the productivity, or rate of turnover. Rapid growth and rapid consumption may lead to low biomass—and a reef does demonstrate a very fast pace of life. A second important point is that because fish are mobile many, including many of the numerically abundant species, roam over much larger areas than just the reef itself; they may travel over adjacent seagrasses and sand beds, or they may pursue shoals of prey into deep ocean water. Fish form major connections between reefs and adjacent habitats.

The complicated ecological structure of fishes on a reef is clarified by the use of nitrogen isotopes to determine the trophic level of any particular fish. The trophic level is a measure of how it feeds, with plants being level 1, herbivores level 2, carnivores level 3, and higher carnivores that eat those carnivores being level 4, etc. You can most easily see what a fish has recently been eating by examining its stomach contents or by simply watching it. However, such observations are usually too brief to show its typical diet. So the ratio of different stable isotopes of nitrogen is used. This changes very slightly with each step along the food chain. Most nitrogen is in the form of $15N$ (nitrogen with 15 protons and neutrons in the nucleus), but $14N$ (with one fewer neutron) forms about 0.4 per cent of nitrogen in air; this ratio changes very slightly with each step in the food chain. The $15N/14N$ ratio increases up the chain such that herbivores have a different ratio from carnivores, for example. Sensitive measurements of the ratio of nitrogen isotopes have been used to determine that many, if not most, fishes' trophic level is not an integer, meaning it is not entirely a herbivore, or a primary or a secondary carnivore, but has a trophic level of, say, 1.5, indicating that it may be mostly a herbivore but consumes small invertebrate animals too, or it may be 3.6, meaning it is a predator of primary carnivores but it also takes in a good proportion of higher level carnivores as well. Some herbivores remain at the same trophic level throughout their life (i.e. eating the same plants), while some carnivores change their diet as they grow larger, feeding on progressively larger carnivores at higher trophic levels. Such measurements give a much more accurate picture of how a coral reef works.

Marked interactions have been seen between fish and the health of coral in all oceans of the world. Some examples are easy to understand: if we eliminate fish which feed on grazing sea urchins, then the sea urchins will increase substantially; urchins graze by rasping on the coral rock, and the cascading effects might include much greater reef degradation. Studies in East Africa and the Caribbean have shown how consumption of triggerfish released pressure on their sea urchin prey and led to an increase in urchin numbers. This led to reductions in algae as well as reef erosion. In the Caribbean, there was a well-researched, massive mortality of urchins from a disease thought to have reached the reef via the Panama Canal, which all but eliminated the dominant grazing sea urchin *Diadema*. The cascade of effects that followed included an increase in seaweed at the expense of corals. The concept of predator release of prey is particularly significant on coral reefs in the present days of overfishing, and the explosion in algae growth that may result could swamp a coral reef. There is nowhere quite like a coral reef for complexity of possible effects to any given change. One example of this is seen with the explosions in numbers of another major coral predator, the crown of thorns starfish.

The colourful crown of thorns starfish is a notorious coral reef predator. It grows up to about 40 centimetres in diameter, with eight to twenty-one arms, covered with poisonous spines, and comes in a variety of striking shades of blue and red. It is a very efficient predator of corals. It was formerly thought to be one species, *Acanthaster plancii*, but is now considered to be a group of at least four similar species separated geographically. Plagues of these starfish have appeared on occasion, always unexpectedly, and in the course of about a month they can consume more or less all the coral over hundreds of hectares of reef. They feed by hunching themselves over a large, relatively flat boulder coral or

15. Crown of thorns starfish in plague numbers, consuming branching corals. Bright (white) corals have already had their tissue totally consumed (coral in right foreground is next!).

by curling themselves around the branch of a staghorn coral, extruding their stomach to dissolve all coral tissue beneath and sucking back the digested contents. Then they move to the next patch. In such plagues, there may be several starfish in each square metre, and plague populations can number tens of millions (Figure 15).

Their natural range extends from the Red Sea in the west through to Panama in the Americas, but they do not occur in the Caribbean. An occasional adult crown of thorns starfish is a natural sight on most coral reefs. However, the earliest known plague outbreaks were first recorded in Japan in 1957 and on the Great Barrier Reef in 1962; outbreaks have been reported increasingly frequently since. For some reason, every now and again they appear in vast numbers, moving on to a reef from deeper water, consuming just about every coral they can get to (and showing some cannibalistic behaviour for good measure). Then, when there is absolutely nothing left to eat they disappear equally suddenly, leaving behind a spectacular white expanse of bare coral skeletons. A week or two later these skeletons become darkened as they are colonized by filamentous algae, and the dead corals become subject to countless burrowing, tunnelling, and boring animals which quickly exploit this newly available habitat.

Recovery by corals from an outbreak takes years. Coupled with other pressures on the reef, these starfish plagues greatly reduce reef stability and persistence, and at times even a reef's very existence as an ecological entity for years to come.

There may be several causes for crown of thorns outbreaks. Increased nutrient flow from sewage or from agricultural land, which provides planktonic food for the animals' early life stages, appears to be one factor, at least in some places. Juveniles remain cryptic and feed on coralline algae initially but switch to corals when young adults. Other postulated causes of population explosions have been overexploitation of some of its predators, including the large Triton gastropod molluscs, much sought after for their attractive shells. It has been suggested that there is a threshold in density that needs to be reached, above which exchange of hormones or signalling chemicals somehow triggers an explosion in numbers and a migration from deeper water onto a reef. It has been shown too that adults down-current of other large groups will follow chemicals excreted by those adults on reefs, so that they are led towards the reef.

Many control methods have been tried after various outbreaks, but in most cases these measures have been too little and too late. Simply cutting up a starfish makes the situation worse because multiple adults can develop from the original. Preferred methods now include collecting them by hand or injecting them with toxic chemicals, but these methods are very labour intensive and inefficient. The fact that they are used at all reflects the urgency, even desperation, to control the damage that a plague can cause. These methods might be most usefully attempted near an economically valuable tourist site, for example, but they have little impact on a wave of starfish travelling along part of the Great Barrier Reef. Predicting plague outbreaks remains difficult too; their larvae, for example, can travel thousands of kilometres.

Fish commensals

The variety of ways different species of fish have evolved to live together or with completely different invertebrates to mutual benefit is countless—and intriguing. Some seem weird to us: certain species of the attractively named pearlfish have a somewhat less attractive habitat of living inside sea cucumbers, entering and exiting through the sea cucumber's anus. Shelter is the main benefit gained, and it seems unlikely that the sea cucumber obtains reciprocal advantage. The relationship between gobies, a type of small fish, and shrimps, which share a burrow, seems rather more appealing to us. The shrimp digs the burrow and, while the shrimp does all the excavation work, the fish stands guard, flicking its tail to warn the shrimp of danger.

But perhaps the most iconic symbiosis on the reef is that between clownfish and sea anemones. In the Indo-Pacific, but not in the Caribbean region, twenty-eight anemone fish species form symbioses with about ten different species of anemone (Figure 16). Some of these fish are specific to one species of anemone while others will adopt any or all species of anemone as a home. Anemone tentacles have stinging cells yet these symbiotic fish avoid being stung and killed. Several mechanisms are postulated: the fish might exude a chemical which suppresses the triggering of the stinging cells, or the constantly flickering, restless activity of the fish might transfer chemicals from the anemone onto the fish's body, so that the fish is sensed as being part of the anemone—after all, one tentacle does not sting another. In return for protection, the anemone benefits from more food, including that from the fish's faeces. The latter is certainly likely to benefit the anemone's zooxanthellae; all anemones showing this

16. Anemone fish, often called clown fish because of their striking coloration, living in a sea anemone in a remarkable symbiotic arrangement.

association contain zooxanthellae too. The presence of anemone fish also increases the proportion of time that the anemone's tentacles are expanded and thus functioning, so there are multiple benefits to both.

A final charismatic example comes from symbiosis between different species of fish. At 'cleaner stations' on a reef, larger predatory fish line up to have any parasites infesting their skin, gills, and even their mouths removed by cleaner fish. Such cleaning stations are commonly at fixed locations on a reef, where a fish signals its desire to be cleaned by adopting a special posture, an invitation to receive attention from the cleaner. The little cleaner fish restrict their diet to dead tissue or parasites, and the predators do not eat the cleaners. However, rarely is a trick missed on the reef: another fish which looks like the cleaner, with the unlikely name of sabre-toothed blenny, *Aspidontus taeniatus*, is a cleaner mimic which will similarly approach the large fish waiting at the station, but instead of helpfully cleaning it, this blenny will take a chunk out of the larger fish's tissues, such as its gills. Naturally, this deters the large predator. It provides a fascinating example of population balancing because the number of sabre-tooth blennies can never be too high or else no large fish would come to be cleaned again.

Reef fish biomass, productivity, and fishing

Given the importance of reef fishes to so many human communities, an important question is: what is the productivity of the reef habitat in terms of fish, or, how many fish can be extracted without compromising the system? Because reefs have been already so heavily fished, it is difficult to obtain accurate estimates of what an untouched reef might actually look like. One measure for doing so is reef fish biomass.

In one of the least affluent regions of the earth, the Western Indian Ocean, there have been some areas that have been protected from fishing for several decades and it has been suggested that the reef fish biomass there might reflect the true level of fish that a well-managed reef might be able to support. But a recent study of historical fish catches between the 8th and 14th centuries, carried out by examining Swahili kitchen middens, has shown that the species' composition several centuries ago cannot have been the same as it is today. Even in relatively well-protected reef areas in Kenya, for example, the size of adjacent human populations is such that reef fish composition has drastically changed over time. Although those sites are protected now, some of the best-managed reef areas still show marked changes and depletions of fish populations, so that they cannot really tell us what the fish population in a 'pristine' reef might be. How much fish might have existed under the best hypothetical conditions millennia ago?

Several researchers have assessed reef fish biomass in a standardized way across the world, focusing on the potentially very rich zone at 9–10 metres deep. Most places measured have a biomass of reef fish of about 1,000 kilograms per hectare or less. In the case of particularly mismanaged reefs, there may be only about 200–400 kilograms of fish per hectare. The best-managed marine protected areas in the Indian Ocean with people nearby have a biomass rarely more than about 1,500 kilograms per hectare. In many countries, areas of reef that are designated as being protected in fact showed no greater fish biomass than designated fishing areas, which suggests that such areas are protected in name only. The same situation holds across the Pacific as well. Only very few locations in the world have been found where the reef fish biomass dwarfs that of even the best-managed protected areas near inhabited places. Richest of all is the mostly uninhabited Chagos Archipelago which had not been heavily fished for many years, and where the biomass at 9 metres depth was over 7,500 kilograms per hectare, six times that of the largest biomass found elsewhere, and a hundred times greater than the most overexploited reef systems in this ocean. Similarly high biomass has been found in the other two major ocean basins also: in Kingman atoll in the Pacific and Cozumel on the Caribbean side of Mexico. Tellingly, in both these places fishing is very low too. In the Hawaiian island chain, biomass values were around 500–1,000 kg per hectare in inhabited islands, rising to 3,000, and in one particularly remote area to 4,500 kilograms per hectare, the higher biomass locations having the most apex predators (like sharks or groupers) of all the locations measured.

There are differences between the three sites with the highest biomass in terms of the kinds of fish that make up the biomass. Kingman and the richest Hawaiian site, for example, have a very high biomass of sharks, much more so than Chagos, which has a higher biomass overall. In any case, the message is depressingly clear: where there is any fishing at all the fish biomass goes down severely; it takes very little fishing effort to cream off the riches.

One aspect of fish physiology is crucially important to replenishment of many fish species: the larger an adult female fish is, the more eggs it will produce. This is possibly an obvious point, but, importantly, the increase in egg production is not linear. To use a hypothetical example, a single 10 kg female might produce many millions of eggs per year, while ten 1 kg females of the same species would produce only a few thousand per year. If we remember that the larger fish are the most prized, we can immediately see that the damage done to the ecosystem by removing the largest fish is exponentially greater. I would stress here that no blame should be attached to those fishing at subsistence level for collecting what they can; for these people, it is usually a matter of survival. Leaving this consideration to one side, it is nevertheless a salutary point to note that even a very modest level of fishing intensity can cause much ecosystem distortion very quickly. In areas that were once protected but which then permitted fishing, ecosystem collapse happened in only a very few weeks.

Fishing is an emotive subject because of consideration of the needs of those who live at subsistence level, but we must understand the situation thoroughly if we are to try to do anything about the overfishing problem. It is demonstrably no good claiming that local people fish sustainably, as is so often claimed. I have challenged many people to produce examples of where an active coral reef fishery can sustain a high biomass of, say, over 5,000 kg per hectare. On the other hand, there are countless examples of 'sustainable' fisheries that sustain only a greatly depleted level. The importance of restoring fish abundance and productivity is growing now that the issue of food security has become so prominent.

Some of the above studies were carried out partly because fishing is thought to be one of the most pervasive threats to coral reef maintenance, due to the important function fish have in regard to the coral reef itself. Because fishing the reefs is also essential to many people's survival and nutrition it has been accepted as being an essential pursuit, a necessary evil, as it were. There is, as a consequence, an exploitation gap: the gap between the high biomass/productivity areas we see, for example in Chagos, Kingman, and part of Cozumel, and the low biomass areas (which lack large, reproducing adult fish) nearly everywhere else. It is a curious fact that, as far as I know, almost no area has been shown to have fish biomass levels of between 2,500 and about 5,000 kilograms per hectare. With, to my knowledge, one recorded exception in the Hawaiian chain, there are only those with the higher biomass figures, which are now very rare, and the almost universally low biomass figures seen where reefs are exploited, which is almost everywhere else.

Chapter 7
Regional scale pressures on reefs

A fundamental character of a coral reef is the enormous number of interlocking connections between components, a consequence of extremely high biodiversity. As the number of species in any ecosystem increases, so does the number of connections between them, and connections rise at a rate much faster than the number of species. Connections can be destroyed, as can the species themselves. The question of whether a reef ecosystem is more robust to outside pressures as a result of this huge number of species and connections, or made fragile because of them, has been endlessly discussed. The answer must be 'both', depending on what aspect of the reef system is being discussed, its geographical location, and what the pressures are.

Reefs can resist being damaged to a certain extent. 'Redundancy' is a particularly important concept. As an analogy, a complex machine like an airliner has key parts and systems duplicated, so the airliner can survive the failure of one. A reef can be viewed similarly. For example, the Caribbean parrotfish *Sparisoma viride* is one of the most important grazers and, where it is overfished, we find that there is no other fish species that adequately takes its place, especially when invertebrate grazing species such as urchins have also been removed; algae cover everything and the reef degrades. There is no redundancy for this 'grazing system', the reef's integrity is threatened and the reef is fragile in this respect.

However, in an atoll lagoon in the Indian or Pacific Oceans, large areas of stagshorn coral can be destroyed by disease for example, and will likely be replaced fairly quickly by one of a dozen other species of stagshorn coral. That reef continues being a coral reef. There is high redundancy for this component of the system so for this component the reef is robust. Different kinds of impacts have different consequences. Further, an impact that kills a reef can be different from the one that prevents its later recovery, so both aspects require redundancy for a reef to survive impacts.

However, the pressures are varied and are increasing, to the detriment of reefs generally, and this is of great importance to humankind. For millennia reefs have supplied food, coastal protection, and shoreline stabilization, joined recently by pharmaceuticals and revenue from tourism that moves money from wealthy places to some of the poorest. Many countries are composed entirely of reefs while many more are partly protected by them, and a number of these are heavily populated. Throughout the Quaternary Period glaciations have come and gone, and reefs have survived them. The time since the end of the last glaciation is known as the Holocene but recently it has been proposed that the time from which human impact on the environment has become significant should be renamed the Anthropocene, because humans are causing geological-scale impacts. We have, for example, now moved more surface sediments than did all of the Earth's rivers and glaciers during the last ice age.

There is some disagreement about when the Anthropocene should start, perhaps from the beginning of the Industrial Revolution, or perhaps as recently as only a few decades ago when our climate started changing noticeably. Some have said it should start two millennia ago, because already substantial changes were taking place as great ancient cities developed, farmed the land, and then were abandoned when their populations outgrew the ability of accessible habitats to support them. If this seems exaggerated we need look no further than the huge accumulations of turtle and mollusc shells piled deep in many coastal locations such as Turtle Mound in Florida (actually about 30,000 cubic metres of oyster shells) and huge turtle shell middens in several Arabian countries such as Oman and some islands in the Red Sea, testament to the enormous quantities of seafood that were being extracted thousands of years ago.

Whenever the Anthropocene started, the point is that changes being wrought by humans on reefs now match or exceed anything that reefs have experienced during the preceding several thousand years, and they are changing. Many other habitats are now in a similar position, but coral reefs have for a long time been considered to be the litmus, the canary in the coal mine, which foretells substantial and deleterious changes happening to the Earth system. They are a major ecosystem that is suffering unwise use to an extent that now threatens their survival. This is not an exaggeration: over a third of reefs that existed 50–100 years ago are now largely dead, and another quarter or even a third (depending on the region) are in critical condition; only about a quarter remain in a healthy state.

There are various causes for the decline of reefs, and they can be grouped into two main categories: those caused by numerous, usually localized, impacts, such as different forms of pollution, overfishing, and shoreline developments that create large quantities of sediment; and those caused by the more recently recognized and probably more serious, long-term factors associated with climate change. All of these are ultimately interlinked and all are caused by human activities. Further, many of these factors occur concurrently, each one exacerbating the harmful effects of others. It would be bad enough if these effects were simply additive, but their effects in combination are commonly multiplied: their effects may be synergistic.

Local pressures

Sewage and run-off. A major local effect has long been that of nutrient enrichment. Coral reefs are well adapted to living in nutrient poor conditions, but sewage, and run-off from fertilized farms, adds much inorganic matter, especially compounds of nitrogen and phosphorus that fertilize plants, which leads to vigorously increased growth of fleshy algae. Unfortunately, these algae commonly include species unpalatable to grazers, so algae thrive and out-compete the much slower growing corals and soft corals. Seaweeds take up space, grow faster than corals, and shade them. Over the past fifty years, countless reefs have been destroyed this way, so that many former reefs are now merely limestone platforms covered with algae and devoid of corals. Furthermore, such pollution can also inhibit corals directly, because phosphate has an inhibiting effect on the coral's calcification process. Phosphate concentrations above just 1 micromolar solution will reduce its calcification and growth.

The stimulation of plankton is equally important: more plankton in the water means that less light can penetrate to the seabed for photosynthesis. Other linkages are being discovered as well: the crown of thorns starfish discussed earlier is a case in point, in that increased nutrients, whether from sewage or agricultural run-off from fertilized fields, stimulate phytoplankton growth, which may increase survival of this predator's larvae.

Landfill, dredging, and sediments. Landfill and dredging have localized but extremely severe effects on coral reefs. At its most obvious level, landfill simply buries the reef, and there are many examples where a continuous fringing reef, or perhaps a series of offshore patch reefs, is now buried under concrete. Fringing reefs are flat, very shallow, and by the seaside, so they provide highly desirable real estate when built upon. It is an unfortunate fact that the price (not value!) of a hectare of coastal reef increases enormously when it is no longer reef (the same also applies to seagrass beds and mangroves). This is not whimsical: about half of the Arabian side of the Arabian Gulf is now artificial, with enormous consequent loss of all marine habitats including coral reefs, seagrass beds, and mangroves, all of which are nursery grounds for marine species. Many other parts of the world are being similarly developed.

One major problem, which exacerbates this, is that of perception and economics. A simple matter like terminology is very important and allows coastal engineers, for example, to 'reclaim' a reef when in fact they are not 'reclaiming' anything at all but 'claiming' an existing marine ecosystem. The terms 'landfill', or 'reef-fill', would be more accurate. Underlying this problem is simplistic economics. The tyranny of the spreadsheet means that a development company, for example, or a harbour authority, costs only its own immediate activities—their own labour, concrete, and equipment used to convert living habitat to concrete building foundation (they would use the term 'develop' rather than 'convert', of course). What are omitted from the calculation are the wider costs, such as maintaining fish nursery grounds and an adequate food supply for marginalized villagers, or the cost of removing shoreline protection provided free by a healthy reef, which is something particularly important to many tropical countries. Vast areas of land are now being created by landfill in many tropical countries. One colleague from Bahrain describes that country as being the only island that no longer has a coastline, meaning that almost none of it is left in its natural state. The trouble is, humans do need that natural state for many different 'services'.

Such landfill commonly has a triply bad effect because material used may be taken from adjacent habitats (commonly misnamed a 'borrow pit', without the usual sense that 'borrowing' has of 'returning it later'), thus the adjacent habitat is destroyed as well. The resulting reef burial has been called 'soil improvement', which it is from the limited perspective of the constructor. Furthermore, while this is going on sediment plumes invariably can be seen flowing several kilometres down-current of the dredging, smothering yet more distant habitats with silt. Corals are particularly sensitive to sedimentation. Pollution control methods for such activities do exist, such as deployment of screens around cutting heads and dredging units, but they are expensive and usually the first kind of activity to be sacrificed to cost saving. The so-called footprint of such tropical shoreline-altering activities invariably extends far beyond the development itself, even though, today, most such developments are supposed to comply with some kind of environmental impact assessment and damage mitigation procedures.

It is not easy to quantify effects of sediments on corals. Responsible engineers are keen to find a simple measurement of what will cause harm to a reef, but this is difficult because there may be several hundred species of corals and soft corals making up the reef, each with widely varying susceptibility to sediment and with different responses to light reduction caused by suspended sediment. The two impacts act synergistically. Sediment can be actively shed to a certain extent by most corals but this requires energy expenditure, which comes essentially from its photosynthesis just when the shading of light by sediment in the water is greatest. Colony shape is important too: open lattice branched forms do not accumulate sediment so suffer less than, for example, leafy forms. We know that the water bathing a reef is usually clear, with corals typically adapted to about 5 milligrams of sediment per litre or less. However, shoreline excavation may increase this tenfold; and a mere fourfold increase can cause mortality in many species of corals. While all species can tolerate brief exposures to high levels, such as might be caused by natural events like storms, prolonged exposure to greatly raised levels for the duration of a major construction event has often killed several square kilometres of coral reef.

There are many more locally significant impacts of this type. Anchors from boats cause heavy destruction, and in some popular recreational cruising grounds, such as in the Caribbean, anchoring has destroyed entire reefs along moderately sized islands. Quite simply, the anchors drag, and even when they don't, the boat swings about its chain, grinding down the corals into rubble. Even small recreational yachts can kill a circle of coral reef several tens of square metres in size, with each anchor drop. The provision of mooring buoys would provide a simple solution to this problem, but their use is far from universal.

Chemicals and metals. The range of industrially derived chemicals discharged into the sea is growing all the time. Many of these chemicals, organic compounds and several metals used in industrial processes, are directly and highly toxic. However, even when their levels are very low, they can accumulate higher up the food chain so that top predators may be killed or suffer physiological impairments. Effects range from loss of fertility, slowed growth, and, in some cases, more bizarre effects like sex change. Some are accumulated in fatty tissues and others might affect neuronal systems or liver function, for example. The range of effects is enormous and this subject has been extensively researched for many years. It does seem, sadly, that very often this research does not lead to clear controls or changes in legislation in many places, and too often work on this has been little more than a post-mortem, carried out to see why a reef system (or other marine habitat) was killed.

Pesticides may have disproportionately large effects. These chemicals are designed to kill, so severe effects should not be a surprise. In reefs some pesticides inhibit zooxanthellae, others affect the animal part, and some affect all life. The main source of pesticides has long been the run-off from nearby agricultural land that has been treated with these substances. New and increasing sources of pesticide are from fish farms and shrimp ponds, where biocides of various kinds are applied vigorously to prevent disease transmission amongst the unnaturally high densities of animals that are kept in such enclosures; many of these biocides then leak on to nearby reefs.

There are many exotic and sometimes unexpected effects from chemicals. Harbours and marinas are a source of significant quantities of toxins from antifouling paints. These chemicals are designed to kill marine life with the intention of preventing fouling on boat hulls so, not unexpectedly, they do exactly that. One recent example is a 'booster biocide' which inhibits the photosynthesis of zooxanthellae in corals at exceptionally low concentrations of only 50 nanograms per litre, or 50 parts per billion. Naturally, it makes very effective antifouling paint.

The problem of pollutants is compounded with that of sediments noted earlier. Sediment particles adsorb toxic compounds onto their surfaces and concentrate them. Filter feeding animals which feed on these particles therefore not only ingest the sediment particle instead of a food particle, but also take in a packet of concentrated, adsorbed toxic substances as well. This is why many beds of sediments that were so contaminated from activities a century ago now must not be disturbed because of the risk to adjacent marine life, or else need to be very carefully removed.

Oil. Oil pollution usually hits headlines when it happens, but oil has complicated effects on coral reefs. It is commonly held that oil and water don't mix, but that is only partly true. Many substances in oil are water soluble, and some are highly toxic to most marine life. Arabian Gulf oil is a fairly light crude, and the many oil spills that occur in the Middle East cause limited damage; the oil simply floats over the top of submerged reefs without doing much harm. (This is not to be confused with the effects of any kind of oil on shorelines, where it accumulates, leaving thick tar.) This oil can be contrasted with some Caribbean oils, which have notoriously damaging effects on coral reefs. In one famous oil spill in Panama near the Smithsonian's Galeta Island research station, mortality was very heavy and extended many metres below the ocean surface. The difference in the proportion and identity of some of the organic substances in oil, especially water-soluble components, lies at the heart of different oil toxicities.

Frequently, pollutants are discharged in pulses, so that, although the average quantity over a year might have been tolerated by coral reef organisms, massive, episodic pulses commonly tip them over the edge. Water conditions after a pulse may quickly return to normal, but the damage will have already been done.

Nuclear explosions. Then there were nuclear weapons tests. From the mid-1940s to the early 1960s many weapons were tested on certain atolls in the Pacific. Atolls were an extraordinarily bad choice for such testing since they are amongst the most diverse ecosystems on the planet. The surface and sub-surface explosions heated surface water to 55,000°C and created shockwaves 30 metres high, with shock columns extending to 70 metres deep in the lagoon, travelling at 80 metres per second. Entire small coral islands vanished, and coral fragments were flung onto the ships observing the explosions. In Bikini Atoll, where considerable work had previously been done on the taxonomy of its corals, a survey done fifty years later, in 2005, showed that more than a fifth of the original coral species had not yet recolonized from other areas.

Spectacular for its scale and destructive effects on atoll-sized reef systems, weapons testing on atolls eventually did stop, although that had little to do with the biological richness of the sites.

Coral diseases

In the last few decades many coral reef diseases have been identified. It is difficult to culture most of the disease organisms—which is the usual first step in identifying any disease—so it has rarely been possible to determine exactly what many of the pathogens are. As a result, some researchers have used the word 'disease' only when the pathogen is known, but 'syndrome' when it is not. A syndrome may become a disease when research determines the identity of the pathogen.

Probably the first documented coral reef disease was of Caribbean sponges in 1938 when 70 per cent of individuals in examined areas died from a fungal infection. Since then, reports of disease have increased. The increase in incidence of disease has been linked to deterioration of water quality. Many diseases and syndromes are associated with multiple infecting organisms: viruses, bacteria, protists, and minute animals. With many diseases, however, it remains unclear which of these might have been the original pathogen and which simply appear immediately afterwards to feed off the newly available necrotic tissue. There is even a tiny crab about a millimetre or two across associated with disease patches on table corals, and even where dozens are seen on a dying coral it is unclear whether the crabs, which are clearly eating the coral tissue, cause the disease or are simply following the band of dead tissue killed by another infectious agent.

Sewage discharges and run-off from farmland with animal faeces contain numerous pathogens, some of which are clearly associated with coral reef syndromes and diseases. Most names of these simply reflect their appearance: white band disease, black band, white pox, yellow blotch, and so on. This seems little more advanced than medicine in the 18th century! White band disease in the Caribbean has had critical and spectacular effects, especially on branching *Acropora* including the giant elkhorn coral, *Acropora palmata*. The white band refers to the denuded and exposed white coral skeleton, and it appears to move because as the disease progresses along the limbs of the coral, it exposes the skeleton, which stays white for a week or two before being overgrown with filamentous algae, darkening it again. Meanwhile the band of exposed white skeleton moves on. The previously vast, impenetrable thickets and forests in shallow areas are not replicated anywhere else. White band disease removed most of these coral colonies in the 1970s.

Some elkhorn coral remains, and the disease itself still exists. The causative agent might be a species of *Vibrio* bacteria, but this is not certain—the epidemic came, killed, and disappeared too quickly for researchers to really find out. A later disease affecting Caribbean corals has been termed white pox disease. It is equally contagious and is manifested by white patches that spread at speeds of up to 2 centimetres per day, with patches merging to kill the entire colony. White pox is caused by a bacterium, *Serratia*, a coliform found in sewage from both animals and people. Whereas white band disease all but eliminated this most important of corals, white pox now keeps populations suppressed. Both diseases also affected thickets of the branching staghorn coral *Acropora cervicornis*.

Caribbean sea fans, or gorgonians, are equally conspicuous on reefs, especially in the Caribbean. Many are the size of a diver, or bigger. Caribbean sea fans have been affected by the fungus *Aspergillus*, which, in the last thirty years, has caused the death of increasing numbers of sea fans. The disease appears as patches of red-brown coloration that expand, darken, and then kill the entire colony. This fungus occurs naturally in air and soil, and its concentrations in rivers have increased enormously with deforestation and agriculture so that now the pathogen is carried throughout the Caribbean from the Orinoco and Amazon rivers, amongst others. A striking link has also been demonstrated with aerial introduction of *Aspergillus* spores from soil swept up over North Africa and deposited in the Caribbean region. The dust has in addition been linked with increasing respiratory diseases in humans. Hundreds of millions of tons of the soil, carrying pathogens, is deposited into the Caribbean during periods that seem to correlate with outbreaks of this disease amongst sea fans.

The sea urchin *Diadema* is widespread, and is one of the most important herbivorous grazers on Caribbean reefs, but those in the Caribbean suffered a spectacular die-back in the 1980s. The first widespread mortality was noticed off Panama, and a wave of mortality then spread out from that point following water currents, so that almost all Caribbean adult *Diadema* were killed within a couple of years. Many juveniles survived, which gave rise to hope for recovery but today, thirty years after the disease, the population densities of this key grazer are still only on average 12 per cent of what they were before the die-off. The loss of these sea urchins was likely to have been a cause of, or at least a major contributor to, the increase in algae now found throughout Caribbean coral reefs. Increasing nutrients from sewage and farmland, coupled with overfishing of herbivores, were important too; it may be futile to try to determine which factor is the most important because it is clear that all of them have acted in concert.

Caribbean reefs appear to suffer more from disease than is evident in the vastly larger Indo-Pacific, which may be partly because the Caribbean basin is very heavily used by people. But diseases are also known and are increasingly researched in the Indian and Pacific oceans too. There have not been such widespread outbreaks of disease in these oceans, perhaps because of the more dispersed nature of the reefs, though this is speculative and the Indo-Pacific has experienced vast and equally destructive plagues of other predatory animals, notably the crown of thorns starfish.

I have explained at some length the pollution and disease problems of coral reefs for two reasons. First, reefs support a greater diversity than anywhere else in the oceans. They also support, through their potentially very high productivity, greater numbers of people per unit area than any other natural ecosystem. The severity of impact by humans on reefs has a disproportionately high repercussion on us. Obviously it is true that, for example, deforestation of several hundred square kilometres of Amazon rainforest greatly affects the local people there, but the extent and consequences of destruction of coral reefs can be argued, from several standpoints, to be equally or even more severe. Furthermore, reefs are the oceans' early indicator of their environmental state.

One other factor may have severe consequences—it certainly has striking effects—and that is invasive species. These are species that are not indigenous to the area, perhaps introduced in ballast water from tankers, attached to ships' hulls, or by an accidental release from aquaria. Most introduced species don't do much damage, but some find a vacant niche and their numbers increase hugely, at which point they are termed invasive.

One of the most striking examples of all is the Indo-Pacific lionfish, which was introduced into the Caribbean in the early 1990s, probably from aquaria in Florida. Two species have appeared, *Pterois volitans* and *Pterois miles*. This highly venomous and colourful fish has undergone a spectacular explosion in numbers. In just a few years it has spread throughout the Caribbean, living sometimes in huge densities, feeding off Caribbean fish which seem to be naive to its stalking behaviour. Engorged and fatter than those living in better balance with their surroundings in the Indo-Pacific, Caribbean lionfish multiply prodigiously; one female can produce 2 million eggs each year. Hunting competitions, cooking recipes, and culling methods have been tried, but in the end all these are futile for control purposes because the depth range of lionfish is much greater than divers can catch them. Their extent is now too great and they breed too prolifically. They will probably never be eradicated from the Caribbean, and lionfish have been called the most important marine invasion that has occurred on coral reefs.

Coral reef fishing

The vision of the tropical village living sustainably alongside its coral reefs with a man fishing for his family is an idyllic picture, but it is, sadly, largely a myth. There has probably never been a time since overexploitation began in the 18th century, except in rare circumstances, when only a tiny human population fished optimal numbers of fish in a sustainable way, leaving enough mature adults to allow for continued levels of fish in later generations. There are examples of traditional management which maintained, at least in the past, large fish stocks, but these were, for example, in royal preserves where the penalty for poaching was extremely severe. Today, where there are high concentrations of people, depletion of reef fish can be swift and extreme, disrupting the function of the reef, and thus its continuing ability to support human communities.

Today, over half of the world's coral reef fisheries are overexploited. Total landings of reef fish are two-thirds greater than can be sustained. Put another way, the area of reef required for fishing is much larger than the area actually available, and several more reefs the size of Australia's Great Barrier Reef would be required to sustainably support the existing catch. Furthermore, many, perhaps most, reported catches of coral reef fisheries grossly under-record the true catch. Coral reef fisheries themselves are only a small fraction of global fisheries, but they support many millions of people.

The problem escalates when increasingly desperate people try to catch the diminishing numbers of fish in more and more destructive ways. Nets become snagged and lost, causing what is termed 'ghost fishing' for years or decades, and explosives may be used including those made from easily available fertilizers. Poisons including cyanide and pesticides such as DDT are sometimes used. Commonly, large areas of reef are irreparably destroyed in the process and profits can be large. It was estimated a decade ago that dynamite fishermen in one Asian country earned three times more than a local university professor, and many times more than the easily bribed local coastguard or policeman. When one area was wiped out, the poachers simply moved on.

We saw that unfished reefs might have an average of well over 7,000 kg of fish per hectare while fish biomass on exploited reefs is commonly only one-tenth of that, sometimes far less. Further, we saw that there is an almost complete absence of examples where fish exist in a biomass in the region of about 2,500 to 7,000 kilograms per hectare, with most reefs today having biomass of about 1,000 kilograms per hectare or less. Possibly such reefs have just never been found, or perhaps when a new reef system is discovered and is exploited, fish biomass traverses that gap—the 'exploitation gap'—very rapidly as the easiest components are caught and fish biomass plummets from the rich to the low state. There comes a point on many reefs where trying to fish becomes an increasingly unrewarding endeavour.

Where fish are removed from the very complex food web, a cascading collapse in other parts of the ecosystem can be seen, to the detriment of the whole reef. This has been demonstrated countless times, and we need think no further than the removal of herbivorous fish, so that algal growth is no longer suppressed. Add in some sewage to encourage more algal growth, and the result is usually a complete domination of algae at the expense of corals and habitat complexity.

The relentless downward spiral of the ecosystem follows, explained by analogy with the Ponzi effect. Charles Ponzi conducted a famous financial scam in the 1920s, whereby people invested in his scheme with the promise of high interest on their capital. Large returns were made, but the money used to pay them was not from interest earned but from the capital given by later investors. There comes a point when such schemes inevitably unravel. What has this to do with exploiting reef fish? There is no suggestion of any fraud, but the principle is as follows: the lowest level of fishing exploitation might be a man in a boat, fishing for his family. To make life easier (nobody can blame him) he purchases a bigger boat and engine. How does he pay for it? From the capital provided by the reef, namely from selling more fish; he then has to catch more fish every day just to pay for it, before he has any left over to eat. He forms a consortium in which people are employed to process this larger quantity of fish. He pays their wages by catching more fish (more reef capital). The village funds a small pier, then a larger harbour of the sort seen all around the tropical world, then perhaps a freezer plant. So, more and more fish are taken to pay for it all. The biggest fish have the greatest monetary value; they literally provide the most capital, yet these are also the fish that also would have produced the greatest quantity of eggs (interest) by a great margin, which would produce the next generation of fish. Thus, the coastal society is not living off 'interest' provided by the reef, but is living off and depleting its capital.

The tragedy is that an economist will talk of 'investing' in fisheries—a freezer plant, say—but it is not an investment at all, it is a drawdown of reef capital. Interest could be possible, but only if fish can be sustained at high enough levels to support an equally abundant next generation, which does not appear to happen. Such 'Ponzi fishing' is very common.

This chain of events is rooted in natural human behaviour. Nobody should blame local communities for wanting an easier or better lifestyle for themselves or their families—and this makes the problem all but intractable. It is Hardin's 'Tragedy of the Commons' for coral reefs.

A second key misunderstanding is use of the word 'sustainable'. Sloppy use of this grossly overused word is seen in any aid or development document today. Biomass in fished reefs is a small fraction of what it could be and, worse still, it is indeed sustained in that depleted state for decades at a time. 'Suppressed' is surely what is meant. Yet, so urgent and intractable is the problem, that phrases such as 'sustainable intensification' are now being introduced (without apparent embarrassment!), which to any practising ecologist is an oxymoron.

Another deleterious effect of fishing is its impact on nutrients. We have seen that it is only tight nutrient cycling that permits such a vibrant system to live in the nutrient poor waters of a reef, but fishing removes biomass, which equates also to removing nutrients. Catching fish, those at the top of the food chain especially, reduces the return of nutrients to the reef system and greatly changes the nutrient balance on a reef.

The challenge is always more difficult because pressure is greatest on the largest, most prized fish. Not only is the biomass (and nutrient) removal greatest, but so is the loss of egg production. Reef fishing started having severe impacts on coral reefs before any other human impacts. 'Management' of fisheries has usually been lamentable, that on coral reefs most of all, as it needs only a handful of fishers to distort the fish balance on a typical atoll. In the words of the eminent fisheries biologist Daniel Pauly, 'We still need to invent sustainability' with respect to fishing. The difficulty is that it is so easy to remove most fish biomass very quickly and simply.

It should be remembered that, in any case, we cannot 'manage' fish levels, as many planners seem to think. We can only manage the human behaviour affecting them. The issue is not one for science so much as for sociologists, or politicians. The problem underlying all fisheries is that provision of food is involved—and food is not an optional extra. There may be a way around the dilemma, discussed in Chapter 9.

Chapter 8
Global scale pressures on reefs—climate change

Climate is changing. Most importantly for this story, it is getting warmer because more heat from the Sun is being trapped by the atmosphere to warm the Earth through the greenhouse effect. This is primarily due to build-up of the greenhouse gas CO_2 mainly from burning fossil fuels. There is a significant upward trend in temperature. Superimposed on this 'background' rise, some years see much warmer temperature pulses than others. In the sea these are termed 'ocean heatwaves'.

Some confusion (and much fuel for climate change deniers!) has arisen because of two things: local, daily, or weekly changes in weather are separate from longer-term 'climate' as the word is used in science; and the Earth has many huge, different thermal compartments that exchange heat with each other, for example the atmosphere with the vast volumes of cool, deep water that periodically rise to the surface. There was a slowdown in the rate of temperature rise after 1998 for a few years, caused by fluctuations in these heat exchanges, and also from increased volcanic aerosols from Mount Pinatubo at that time which reflected sunlight, and a weaker solar cycle, but since then warming has continued. About 93 per cent of trapped solar heat is stored in the ocean. Figure 17 shows the amount that this has climbed since 1960, and all that excess heat has warmed surface

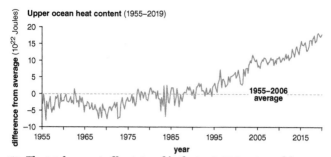

17. The total amount of heat stored in the top 2,000 metres of the ocean, from 1955 to 2019. The horizontal line at the zero x axis is the average between 1955 and 2006.

water by about 1°C. Water has a much greater thermal capacity than the air. Had there been no ocean to absorb the heat, the air would have warmed by 36°C, more than enough to snuff out all life larger than some microbes. The heat that has entered the oceans in the last fifty years is about 300 ZettaJoules, an impossibly high number to visualize, involving 21 zeros. It may be easier to imagine five Hiroshima-size bombs exploding in the ocean every second since 1950—but perhaps not much easier! The rate of rise is increasing too. This is mainly because of rising atmospheric CO_2. It is the biggest problem that has ever hit reefs.

Warming and coral bleaching

Reefs are more affected by the damaging consequences of climate change than most ecosystems. This is not a story for the future, but one that is happening now. The damaging impacts of climate change synergistically multiply the harmful effects already noted. In this respect, coral reefs foretell the global condition perhaps more than other ecosystems. Whereas diseases and pollution have had enormous impacts on many reefs of the world, all are in a sense fairly local. These have now been joined by global effects of warming.

Coral bleaching was the first manifestation of climate change affecting reefs, although initially many people did not recognize it as such. Vast expanses of corals, commonly extending across the whole reef, simply turned white. First researched on the Pacific side of Panama in the 1980s, it was then observed to a massive extent throughout the Indian Ocean in 1998, in some parts of the Caribbean around this time too but mostly later in 2005, and in various parts of the Pacific and Australian region in a complex pattern since then. The phenomenon had been reported in the 1870s but the main causes then were probably sewage and pollution. For a long time, climate effects stayed under most scientists' radar, and even scientific reviews commenting on reef decline written as late as the 1990s scarcely mentioned it. One influential review on potential vulnerability of coral reefs noted, incorrectly, that remote oceanic atolls were amongst the most immune from degradation, but massive and widespread ocean warming has equally affected those distant reefs. Views had to change rapidly.

The first wide-scale bleaching event occurred in 1998, when corals on reefs in very widespread areas turned white, following which corals and soft corals died. Mortality extended to various depths, sometimes to as deep as 40 metres. We realized then that the main environmental driver affecting reefs had shifted to something altogether different. Mortality and later recovery were measured and recorded from that date in several locations, and there have been further smaller warming events since then, particularly in 2005 in the Caribbean. Then, more recently, in 2015 and 2016 there was a particularly severe pair of ocean heatwaves. These caused extensive damage over large swathes of the Great Barrier Reef that, until then, had by and large escaped the worst impacts from the earlier warming pulses. Hundreds of square kilometres of reef were severely bleached, and then killed.

The term 'warming', however, is usually shorthand for several different variables. True, warming is the most important, but there can also be a change in light intensity received by symbiotic algae. Changes to wind conditions affect the sea's surface, so that in the Indian Ocean, for example, the sea remained glassily calm for long periods. As any underwater photographer knows, the brightness at any particular depth is different depending on whether it is stormy above or flat calm, a calm surface letting through appreciably more light. The greater light is sufficient to 'overload' the photosynthetic pathways in the corals' zooxanthellae.

Through thousands of years, corals in any area have become acclimatized to a fairly stable ambient seawater temperature and light levels, and to the range of variation of each. Corals have adapted to certain conditions, and their optimum temperature is only a couple of degrees cooler than temperatures that kill them. With warming and more intense light, the symbiotic algae died. They were expelled, and then the corals died a few weeks later.

The concept of 'degree heating weeks' has proved very useful here. This is the number reached when we multiply the level of temperature rise of water above what corals were 'used to', by the number of weeks it remains high. Different species are, unsurprisingly, affected differently, but the number 8, resulting, for example, from four weeks at a temperature of 2°C above what is usual for that time of year, will generally prove fatal for most corals.

The coral may not die after bleaching. If the conditions are not too intense and do not last too long then re-acquisition of algal cells may allow the coral to recover. But if it stays too warm for too long, re-acquisition of algal symbionts does not happen and the coral will die. The visual bleaching arises because coral tissue itself is usually fairly transparent, and expulsion of the zooxanthellae means the white limestone beneath is seen. Dead coral tissue will slough off, again revealing pure white limestone. However, the bleached appearance does not last more than a couple of weeks because small filamentous algae and many other groups quickly colonize the newly available bare substrate, turning it dark again.

We know that, in general, sea surface temperatures began to rise in the 1970s. But for a decade or two the change was small and could have been due to random fluctuations. After all, seasonal changes tend to be much greater than the underlying upward pattern; the 'noise' in the data is bigger than the 'signal'. By the end of the last millennium, however, the underlying trend of rising temperature, which by then was being vigorously researched, became clear. The trend continues: almost all of the warmest years of the last several thousand years have occurred in the last couple of decades, and this is both because of the gradual upward background trend in temperatures and periodic ocean heatwaves superimposed upon that.

Work on predictions of temperature in the near future is progressing, and all models forecast that the background rising trend will continue and that ocean heatwaves will become stronger, and more frequent, leaving corals less and less time to recover between episodes. Bleaching and mortality could happen as frequently as every one or two years on most coral reefs in the next couple of decades. Given that a young coral needs to be about five years old, on average, before it can reproduce, bleaching events of this frequency will kill corals that are still too young to have reproduced.

The effect on corals has been measured now in many places. One of the longest single time-series comes from the Chagos Archipelago in the central Indian Ocean (Figure 18), which shows a decline in coral cover over fifty years. The decline has been 'stepped', each step being correlated with particularly severe ocean heatwaves. Averaged declining curves in corals from several sites over larger areas have been shown for the Caribbean, Pacific islands, and Great Barrier Reef. As a very rough guide, a cover of about 10 per cent is needed for a reef to calcify enough to still grow; lower cover leads to reef erosion. Coral cover in Chagos shown in Figure 18 is now around 10 per cent. However, this is an

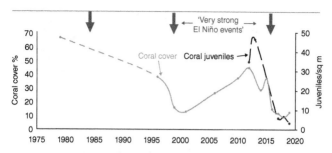

18. **Average coral cover on a group of central Indian Ocean atolls, 1979 to present. Coral cover line on the left is dashed, indicating a lack of observations during that period—the line is unlikely to be smoothly descending. Also shown (large dashes and right *y* axis) is the density of juvenile corals. Data from five atolls of the Chagos Archipelago, ocean-facing slopes and all depths to 25 m deep. Arrows along the top are 'Very Severe' El Niño events (NOAA's Oceanic Niño Index), a measure that reflects ocean warming.**

average value: some sites are higher, and many are considerably lower.

On this reef also, the numbers of juvenile corals settling on the reef have also collapsed to about a tenth of earlier levels. As these juveniles are the next generation of reef-builders, the reduction will have consequences.

A typical reef from this location shows two stages in reef mortality. First, the coral polyps die and their skeletons remain in growth position, but after a couple more years, the skeletons collapse, disintegrate, and then crumble and become swept away, leaving very little structure behind (Figure 19).

Acidification

While warming has been the most important cause of reef destruction during the last few decades, another factor is

19. Top: a field of table corals killed by warming less than a year previously; colonies eroding but still mostly intact. Bottom: reef three years after mortality, when dead corals mostly have become rubble and washed away.

acidification. When CO_2 dissolves, the pH of the water falls. Carbon dioxide dissolves, first forming carbonic acid, and then dissolved bicarbonate and carbonate, all three being in chemical equilibrium. This equilibrium or buffering system works in a rather counter-intuitive way, but, broadly, more acidity reduces the amount of carbonate available to corals for limestone deposition. Some estimates already suggest that coral growth rate has decreased significantly because of this effect. The pH values have dropped by about 0.1 units overall, but this logarithmic scale equates to a 30 per cent increase in the number of hydrogen ions, which crudely means level of acid. (We should really talk of 'reduced alkalinity' rather than acidification because the pH of typical seawater is slightly alkaline: 8.3 is a common value.)

The story is complicated because atmospheric carbon dioxide takes twenty to forty years to reach equilibrium with the oceans—there is a time lag, sometimes called a 'legacy', before carbon dioxide already in the atmosphere has its greatest effect. Taking this into account, it is thought that in twenty or thirty years the number of hydrogen ions in seawater will be three times higher. Our atmospheric blanket is not only increasing but its rate of increase is rising too. Pre-industrial levels of atmospheric carbon dioxide were around 280 parts per million (ppm). Unfortunately, it has been calculated that (given equilibrium with the oceans) a level of 350 ppm would be grossly inhibiting, if not lethal, to coral reef growth. In 2020 it reached about 415 ppm.

Carbon dioxide, then, causes both warming and acidification, and both effects are predicted to continue on a rising and damaging trajectory. Warming is most immediately important, but acidification will take the longest to cause impacts and to be reversed if we could eliminate emissions. Unfortunately, the geological record is not encouraging: after periods of high CO_2 that accompanied most if not all previous global extinction events, recovery time of reefs usually exceeded half a million years.

Sea levels, breakwaters, and hurricanes

Severely damaged reefs erode because most bioeroding organisms on reefs continue to thrive. Their eroding effects might even accelerate with the increased amount of exposed substrate available to the bioeroders.

Coral reefs provide natural breakwaters around thousands of islands and thousands of kilometres of coastline and many governments are showing growing concern at increasing erosion patterns. Shoreline erosion as a consequence of destruction of the reefs is often conflated with the observed rise in sea level, and indeed the two factors act together. Sea level is rising at an average of several millimetres per year, but varies significantly depending on location. The averaged sea level rise in the 20th century due to warming was 17 centimetres and today the rate has increased, partly due to a greater amount of water in the oceans because of melting ice, and partly due to thermal expansion of the warmed seawater. The rise in sea level is causing serious problems for many islands, although the rise by itself would be less important if the coral reefs, which essentially provide offshore breakwaters, could continue to grow. After all, healthy reefs have grown vertically throughout their existence, but in order to do so the corals must be healthy. The combined effect of sea level rise with the reduction of offshore breakwaters means that wave energy striking shorelines is increasing significantly. This is having important consequences for coastal infrastructure such as roads, facilities, and housing in many parts of the world where shores are fringed by the free breakwaters provided by reefs.

Warming is having other effects on the weather too. Hurricane patterns (called cyclones or typhoons in various parts of the world) appear to be changing. These destructive rotating winds are fuelled by the latent heat of evaporation of water. For one to form, the ocean temperature must be greater than about 26.5°C, so they are tropical. But they also need a differential spin imparted to them across their width, a result of different latitudes of the earth moving at different speeds, so they cannot form too near the equator because their north and south sides would be moving at similar speeds; thus they cannot form closer to the equator than about 10° North or South. In the Caribbean, where this has been researched extensively, indices like the Accumulated Cyclone Energy have been developed that combine, size, wind speed, and duration, and these show significant rises over the last few decades. Hurricanes are not becoming more frequent but are becoming larger, longer lasting, and stronger, which leads to more destructive energy overall. Coral reefs that have experienced hurricanes throughout their existence have developed accordingly, but we have little idea how coral reefs will respond to increased levels of hurricane energy. The Atlantic coast of South America, with its uniquely distinctive collection of coral reefs, had never experienced hurricanes until 2004—when Brazil experienced its first.

All this is caused by CO_2 rise: corals are killed by warming; growth of reefs is slowing; more bare surface is exposed for bioeroders to settle and reduce reefs' breakwater effect; sea level is rising; destructive storms are increasing; acidification which reduces growth of surviving corals is increasing; and numbers of new coral larvae are falling. All exacerbate each other, and there are signs that a spiralling, positive feedback condition has already begun to develop, sometimes called a tipping point. The question is: will society rectify this in time?

Chapter 9
Doing something about it

Understanding the main issues

It is helpful, when talking to decision-makers, to use human analogies where possible. Most of these people, who have ultimate responsibility over the fate of coral reefs, have probably never even seen any. So a useful analogy might be disease in humans: we can survive many diseases if we are well fed and healthy, and are infected by them one at a time. But a person is far less likely to survive if afflicted by, for example, measles, malaria, cholera, and influenza all at once—and is malnourished too. Today this analogy can be applied to reefs. Not only are reefs affected by simultaneous consequences of overfishing, pollution, sedimentation, and the rest, they are also stressed by higher than optimal ocean temperatures and acidity levels of the water. This is why about a quarter of reefs are already dead, probably irrecoverably so as far as we can tell, and a further half are affected to various degrees. This leaves only about a quarter that can be considered to be in a good state, with perhaps only a tiny proportion that remotely resemble the state of reefs several decades ago. Continued failure to reduce CO_2 levels means that even if the aspirations of the Paris Climate Agreement are achieved, coral reefs are still very likely to be reduced further by 70–90 per cent by the middle of this century. The threat to coral reefs is existential.

Their decline is not linear, however, and nor is the change in their function as they decline. More commonly, the decline proceeds in steps, coinciding with ocean heatwaves, perhaps superimposed on a decline already caused by one of many 'local' stressors, such as pollution or overfishing. The ecologically stable state of a coral reef is, naturally, presumed to be a condition in which corals dominate. But this is not necessarily the case. We know that there are alternative stable states, as illustrated in Figure 20(a). Imagine that the ideal of a rich, coral-dominated reef is represented by the black ball settled in the valley in the upper sketch of the figure, separated from a lower valley by a hill.

Stress the ball (the reef) a little with a push and (as shown by the arrows in the top ball) it will simply wobble in the valley and settle back into it when the stress is removed. These little pushes represent the normal occurrences that reefs experience daily and annually. Then imagine ecological impacts acting on the reef (the lower of the two sketches of valleys): one might be input of sewage, which stimulates seaweed—this might be represented by a reduction in the height of the ridge—and another might be overfishing, an extra hard shove to the ball. The ball will roll over into the next, lower valley, which in this case represents a limestone platform covered by seaweed. This state seems to be perfectly stable and the ball will not return to the upper valley (the coral dominated state) by any means that we know. Removal of the two stresses is not enough—the area does not easily return to a coral dominated state. This in engineering terms is the hysteresis effect, and it appears to apply to reefs too (Figure 20(b)). Unfortunately, seaweed dominated limestone platforms are becoming increasingly common. It is presumably possible to regain a coral dominated state somehow, but not easily.

Debate continues about details: for example whether seaweed domination occurs primarily as a result of herbivore removal, pollution, or raised temperature killing the corals. The answer

(a) Coral valley   Algal valley

(b)

Coral Reefs

Coral cover

Trajectory if stress applied

stress lifted

Increasing stress

20. (a) Two conceptual diagrams in which the valleys represent different ecological states: coral dominated on the left, algae dominated on the right. In the top example, the black ball is in the coral valley (a healthy reef). Small shoves do not dislodge the ball, which returns to its stable valley. Below: when pollution reduces the height of the valley wall, and another stressor gives the ball a harder shove, it rolls into the lower valley—an algal dominated state. It does not have any known inclination to return back up to the coral dominated valley.

(b) The hysteresis curve is another way to represent shifts in community structure. The top curve shows stability as stress increases showing no change in coral cover until the system collapses. Removing the stressor, however, does not permit the same trajectory back to health, needing much more time before recovery is seen.

114

undoubtedly depends on the specific location. Furthermore, the stressor that changes the ecosystem may be different from the stressor that keeps it in the altered state for decades.

Extremely degraded reefs are now common. The rubble left on many becomes sedimented and covered with a cyanobacterial and algal film. It has been said that the switch to such a state is like moving backwards in evolutionary time, to Precambrian conditions: a 'slippery slope to slime'.

Despite the gloomy prognosis there may be glimmers of hope. Deeper zones on most reefs have escaped some of the worst damage and these can contain reservoirs of corals that could repopulate shallow areas.

Adaptation by corals to warming is a much-hoped-for solution too. We know that many species of coral in naturally warm basins such as the Persian Gulf can survive annual high temperatures of 33–5°C, whereas the same species of corals in the Indian Ocean are killed at 29–30°C. The Arabian corals have clearly adapted—though coral diversity is much lower overall. Whether the adaptation is by the coral host or its algal symbionts is being intensively researched. In a sense it doesn't matter—adaptation has clearly occurred, so perhaps we can assume that it can happen more widely. However, natural adaptation has had several thousand years to take place, and the rate of change seen today might not leave nearly enough time for this adaptation to occur before the reefs die out.

Such hoped-for solutions do rely rather too much on hope. But hope is no way to manage a planet! There are things we can do to effect change, though none will be easy.

Solutions

We must first understand the 'shifting baseline syndrome'. This phrase was coined twenty-five years ago in relation to fishing by

Daniel Pauly, and refers to how our expectation of what a healthy ecosystem should look like changes over time. I have suggested that it applies equally to the condition of coral reefs. Any 'baseline' measured today (and so-called baseline surveys are commonplace) will be markedly different from the baseline that might have been measured long ago. Indeed, there is commonly no easy way of determining what a true baseline condition might once have been. The term 'natural' or 'pristine' has commonly been used to refer to what a particular observer first studied, which might be very different from what it was earlier. A small stress on a reef that might now tip it over the edge into a different phase state might not, perhaps, have had a noticeable effect fifty years ago, but today the stress is acting on a reef already under severe pressure. That reef's 'elastic' has been stretched nearly to breaking point already, so to speak, without us having realized it.

Several groups of solutions offer the most hope, in my opinion, and, almost certainly, all must be used together. A dispassionate assessment shows that the overall damage is already severe, is continuing, with projections showing that further deterioration will happen. The following measures can reverse the slide, but only if governments and authorities get on with it.

Control local damage. At the local scale, we need to control pollution, fishing intensity, and so on—all easy to suggest but very difficult to achieve, particularly in countries which are poor, or have little understanding of the natural systems that support them. These changes require education and management. But we must abandon the mind-set that we can 'manage' a coral reef (or any other marine system) because it is pure hubris to think that we can. What we can manage is human impact on them and, where we can alleviate that effectively, several aspects of reef health will start to repair themselves. The manager who thinks there is no point in, say, controlling fishing or pollution because climate change is going to damage the reef systems anyway needs to know that local actions might make a lot of difference.

Such measures are not hard to put in place—in principle. However, many damaging activities are supported by powerful groups who have vested interests and the backing of well-funded lobbyists. Nonetheless, convincing arguments can be made that taking such measures, and allowing the reef to recover, will lead to the provision of more services to more people—for example, the notable rise of food and fishermen's incomes in areas in the Philippines where this has been done effectively.

Properly applying environmental assessments. These should be made before all major coastal changes and developments and, crucially, the construction industry and oversight authorities need to follow the resulting advice. If done, a large project may have minimal or temporary impact: the country can have its desired development and still retain a healthy reef ecosystem. Sadly, many regions pay only lip service to good advice and have ignored environmental assessments. Equally, many environmental assessments have offered truly inadequate solutions.

Preventing externalizing costs. A company can fulfil its need to make money by making more of whatever it does, but also by getting someone else to pay its costs. Costs here commonly include cleaning up its own pollution. By not doing so, villages down-current pay the cost of resulting damage. In parts of Asia, for example, entire fishing villages have been abandoned when their nearby reefs were killed, along with their reef fish nursery grounds, when up-current construction did not control sedimentation. It is the local people who suffer far more than the developers or the government. Social justice is one of the UN Development Goals too.

Offset schemes. Paying for damage deemed to be unavoidable can be accommodated via environmental or habitat 'offset schemes' or 'carbon credits' and the like. With these, a project incurring unavoidable negative impacts is required to restore some equivalent habitat that was damaged previously. There are of course problems in determining the principles on which the identification of an equivalent habitat, or an equivalent area, may be made, and results have been somewhat arbitrary and difficult to agree. Such approaches may help reefs less than many other habitats, however, because so far it has proved impossible to restore a damaged reef to anything like its original condition. Any increased damage to reefs can only be offset by restoring quite different types of habitat, which is not good from the coral reef perspective.

'Enhancing evolution'. This currently popular subject intends to genetically enhance the ability of corals to withstand warmer temperatures. Whether it involves selective breeding or modern genetic manipulation in corals and their symbiotic algae, some success is being achieved, though at present at levels well below that which will be required. While it is currently impossible to tell whether success will accelerate, performing the research at all is a measure of scientists' desperation to explore all avenues to help corals. 'Assisted gene flow' of the enhanced corals might yet be beneficial on a large and meaningful scale. At present there has been considerable doubt expressed about the possibility of this being conducted at scales that will in fact be important, but the counter-argument is that it is certainly worth trying.

Artificial reefs. Structures, usually made of concrete, have for many years been placed in areas where reefs have been destroyed, to provide a foundation for new coral growth. When these work, they provide a substrate elevated above the old reef where the latter has become a bed of shifting rubble, lethal to newly settling coral larvae. Artificial reefs have had some success in high use areas, but the scale required for any ocean-scale restoration is too great for this to be a universal solution. Also, too often they are deployed where the pollution that killed the original reef is still ongoing, thus almost guaranteeing their failure. Some, made up of hundreds of car wrecks, have been no more than cheap waste disposal, while others, such as some made of car tyres, have later had to be removed when they have done more harm than good.

Marine protected areas. Another group of measures involves developing strictly protected areas. Several agencies and individuals have estimated how much of the reef (or any habitat) should be properly protected. By 'protected area' I mean one where no extraction or destructive use of any kind is permitted, enabling proper refuges and seeding areas to develop. Estimates of the area needed range from 10 per cent to 50 per cent of the total, an alarmingly high proportion perhaps. Special pleading for exemptions, most commonly in favour of continued fishing, are common whenever a marine reserve is proposed but, when these exemptions have been granted, they have almost completely nullified any benefit of a marine reserve. Much resistance to the concept comes from interests wanting to continue fishing, building, or other use that may previously have been unrestricted.

Marine reserves must be chosen and sited carefully. Sometimes a group of several reefs can be selected, with genetic interconnections between them being crucial. The country concerned must have the ability to operate and protect them effectively in the first place—many do not—and there is no point in declaring a protected area adjacent to a large population of people. But protected areas do work if planning is appropriate. Given that we cannot really 'manage' a marine reserve, we must simply allow nature to recover by leaving it to its own devices. This they often do, and numerous examples exist where an effective reef reserve has led to a tripling of protein extraction and fishermen's income, in just three years.

A half-dozen enormous marine protected areas have been declared in the Pacific and Indian Oceans whose remoteness and other circumstances mean that they do not have huge local human pressures, but have an effective governance. These large areas are likely to prove invaluable. Many have the additional benefit that their component reefs are genetically interconnected. They go a long way towards protecting about one-third of the ocean from damage.

Marine spatial planning. This is an evolution of marine reserves. Most countries have some sort of spatial planning on land, where areas are protected, or set aside for industry, farming, housing, etc. Few countries effectively apply spatial planning to the ocean. It involves locating industrial activities as well as protection to different areas. Words such as ecosystem-based, integrated, adaptive, strategic, and participatory are used (not least in UNESCO's definition of the process!), all meaning that use of each area is planned rather than haphazard and often conflicting. The aim is both to protect the environment and to provide the required social outcomes, such as food security, while avoiding unintended damage. Few countries currently apply this effectively.

If there is a significant human population nearby, no amount of control can keep hungry people from impacting it, so, where there are communities too large to permit any hope of conservation, then huge adjacent areas can be zoned for fish farming and food provision, for example. People will obviously continue to try to obtain food from other areas too, so a better zoning approach on a regional and ocean-wide scale will be needed if we are to prevent piecemeal decline throughout the coral seas. Zoning is based on both pragmatism and science, and may run counter to many traditional conservation approaches.

Science trumping 'traditional management'. Very often, traditional ways of doing things are viewed as somehow sacrosanct, with the right to continue in the same way. However, what worked in the past, with small populations, such as gleaning or fishing, simply cannot support the present much higher populations. Also, just because something evolved to become traditional should not be held to mean it is sacrosanct: it is not difficult to think of many traditional activities that are now regarded as highly undesirable—think no further than the ancient and long-lasting tradition of slavery.

Local scale fisheries offer more examples. Only a couple of dozen fishing families can hugely reduce the productivity of a large reef, for example. This is a difficult dilemma—what used to work may no longer achieve the goal of providing fish from a healthy reef.

Nevertheless, it remains essential for local people to have a stake in the system. The principles and the point of conservation must be explained effectively and there must be a willingness amongst the communities to make it work. This has been achieved in an increasing number of places. When this happens, the villagers themselves then have become the best ambassadors for the idea.

A global portfolio of coral reefs. For various reasons, often unknown, several reefs have survived in much better condition than others and have survived the impacts from climate changes better than most have managed to do. These reefs must be identified and given better than average funding for their prolonged survival, including local management of activities on them including protection. Choice of these reefs must be driven by scientific reasons, not political. These reefs might become the seeds of a future recovery once humans have managed to reduce all harmful impacts.

Protection in non-territorial waters. Most of the ocean is 'open ocean'—not covered by national jurisdictions. This stems from the idea of 'freedom of the seas' that emerged during wars between European colonial countries centuries ago. It is now enshrined in the UN Law of the Seas (UNCLOS). Unfortunately, while initially intended to ensure free passage and navigation, it is used also for exploitation, and this does not necessarily come with the required responsibility. Only a few coral reef areas are affected by this as most reefs adjoin land and, unlike the ocean, all land is subject to someone's laws. Some shallow reef areas without islands do exist, and these have seen increasing territorial and resource rights conflicts. At present the requirement for the exploitation of these vast areas to be conducted responsibly is alarmingly inadequate.

Changing the use of former reef areas. Emerging now is the move by some to essentially accept that reefs are one of those ecosystems that will simply disappear, like so many ecosystems have in the past few million years. Those reef areas will not recover, the argument goes, so why not use that shallow, well-illuminated, tropical marine area to best advantage now rather than allow it to slide into becoming a habitat of little use to anyone. Some shallow limestone areas covered with algae, for example, do produce good harvests of fish, albeit no longer coral reef species. This polarizes the views of scientists to some extent. On one hand it is seen as giving people dominion over all other life to use as we see fit—reverting to a biblical tradition perhaps. On the other hand, it is seen as a policy of despair, of giving up, by those whose view is that humans should live more harmoniously within nature. One is utilitarian: what does it matter if this or that group of species disappears when required; the other is more a 'deep green' view. The first can argue that reefs are vanishing anyway, so we may as well control what reef areas become, else they will both be killed and become useless. The deep green view is simple disquiet at the thought of losing something that evolved over millennia to become the ocean's most diverse ecosystem.

The elephant in the room

Carefully avoided in many scientific discussions, government reports, and papers is the issue of human population. Sometimes, even, the subject is ruled to be out of bounds. But rising numbers of people, and their desire for higher standards of living, put increasing demands on natural resources. More people are chasing a fixed or declining stock of reef resources: the area of the planet on which coral reefs can grow is limited, after all. In one sense it is really that simple. Some coastlines have a human population doubling time of only fifteen years, which reflects medical advance and its highly desirable accompaniments such as increased survival of infants. However, this means that current scientifically calculated solutions for a particular section of reef shoreline, for example, are negated when the population doubles. The solution is no longer a scientific one, but has become largely a social and political one. Human numbers are part of the equation, and if we ignore part of an equation then we cannot solve it. Several organizations have found that women's education in poor rural areas and their desire for family planning is a crucial unmet need, which, when it is met, has positive effects on all aspects of the region—from improved human health and welfare, to improved capability of habitats, including reefs, to sustain the communities.

Degraded and overexploited reefs have a very high social cost. We can only make a rough estimate of how many people are malnourished or have died because their local reefs can no longer support them, but the number is many millions. It is difficult to understand why such massive effects remain unknown to most. The issue of food security is paramount today, and dominates many environmental documents. Estimates of mortality of children under 5 from malnutrition are 54 or 56 per cent; such children may at first glance have been killed by a particular disease, but it was malnutrition that made them vulnerable to the disease in the first place. The Food and Agriculture Organization of the UN reported: 'in general the countries that succeeded in reducing hunger were characterized by rapid growth in economic and agricultural sectors. They also exhibited slower population growth.'

Costs and values

It may be commonplace today to draw attention to the difference between the value of something and its price (Oscar Wilde's famous dictum), but the distinction is still overlooked. Average cover by coral on reefs has, for a couple of decades now, been declining at an average rate of 1–2 per cent a year over most of the world. Simple maths shows that the prognosis is terminal unless values are reassessed. As a scientist who has researched reefs for nearly fifty years, I do not want my work to be merely a post-mortem of what reefs once were, but this is all too likely to be the case. I believe it is now a political and social problem to solve, not simply a scientific problem. Otherwise, we would have solved it.

Never has it been more critical to arrest this global trend. At present rates of degradation, coral reefs will probably be the first major ecosystem to become ecologically extinct in this Anthropocene epoch. Even though scientists instinctively show caution in their pronouncements, this possibility of near complete disappearance is being raised increasingly frequently, indicating how concerned we are with present trends. There is also, remember, the 'legacy time' to consider, the 20–40 years' delay described earlier before the full impact of factors such as ocean acidification becomes apparent. The decline of reefs is not a smooth trajectory in any area, of course; it commonly shows periods of stasis interspersed with steps of sharper deterioration. There is the evidence of threshold effects too: only a tiny additional shove is needed to push the black ball of Figure 20(a) into the algal valley once it is already positioned near the top. The ecologist Nancy Knowlton provided a vivid analogy: 'when the thermostat is turned up one notch people tend to expect one notch's worth of additional heat, not a house in flames'. The episodic notches upwards in temperature of the oceans and consequent steps downwards in coral cover on reefs follow this pattern.

It still may be possible to arrest the decline and so continue to receive benefits from the flourishing world of coral reefs, but the political sector is where the change must come from. We know the solutions in scientific terms, some are simple physics and many are uncomplicated biology, but we seem helpless in a political sense. My views here are not new, and I ended a book I wrote thirty years ago with these words that are more urgently applicable today:

> Someone once said in reference to that other tropical wonderland, the rainforest, that its destruction, because of short term negligence and profit, was like stripping a Rembrandt for its canvas. With a coral reef too, its assets are priceless and depend on it being and remaining a mosaic of connected form and living pattern.

We must learn to apply what we know much faster than we have done in the past. Since the first edition of this book seven years ago we have seen further massive ocean heatwaves and another large step downwards in coral reef decline. We know enough to save coral reefs from terminal decline but we have not done so. We must make politicians change as fast as our climate is changing, and as fast as people are stressing this ecosystem, if we are to stay ahead of the game. Many of these stresses affect most other ocean ecosystems too—remember that life on land has never existed without life in the sea, though life existed in the oceans for many millions of years before there was any life on land. People have drawn attention to the fact that many changes today, such as ocean acidification, mimic conditions thought to have existed during past extinction events, though this time humankind is driving the causes. The teeming life of the beautiful and complex ecosystems that are coral reefs can be saved, given the will.

Further reading

Books: both general and fairly specialist

Birkeland, C. (Ed.). 2015. *Coral Reefs in the Anthropocene*. Springer, Dordrecht, 271 pp.

Burke, L., Reytar, K., Spalding, M., and Perry, A. 2011. *Reefs at Risk Revisited*. World Resources Institute, Washington DC, 114 pp.

Côté, I. M., andReynolds, J. D. (Eds). 2006. *Coral Reef Conservation*. Cambridge University Press, Cambridge, 568 pp.

Fabricius, K. E., and Alderslade, P. 2001. *Soft Corals and Sea Fans: A Comprehensive Guide to the Tropical Shallow Water Genera of the Central West Pacific, the Indian Ocean and the Red Sea*. Australian Institute of Marine Science, Townsville, 264 pp.

Gray, W. 1993. *Coral Reefs and Islands: The Natural History of a Threatened Paradise*. David and Charles, Newton Abbot, 192 pp.

Hopley, D. (Ed.). 2011. *Encyclopedia of Modern Coral Reefs: Structure, Form and Process*. Encyclopedia of Earth Sciences Series. Springer, Dordrecht, 1236 pp.

Hopley, D., Smithers, S. G., and Parnell, K. 2007. *The Geomorphology of the Great Barrier Reef: Development, Diversity and Change*. Cambridge University Press, Cambridge, 532 pp.

Hutchings, P. A., Kingsford, M., and Hoegh-Guldberg, O. 2019. *The Great Barrier Reef: Biology, Environment and Management*. 2nd edition. CSIRO Publication, Melbourne, 450 pp.

McCalman, I. 2013. *The Reef: A Passionate History. The Great Barrier Reef from Captain Cook to Climate Change*. Penguin Books Australia, Melbourne, 352 pp.

McClanahan, T., Sheppard, C. R. C., and Obura, D. (Eds). 2000. *Reefs of the Western Indian Ocean: Ecology and Conservation.* Oxford University Press, Oxford, 600 pp.

Mora, C. (Ed.). 2015. *Ecology of Fishes on Coral Reefs.* Cambridge University Press, Cambridge, 388 pp.

Prideaux, B., and Pabel, A. (Eds). 2018. *Coral Reefs: Tourism, Conservation and Management.* Routledge, London, 312 pp.

Riegl, B. M., and Dodge, R. E. (Series Eds). Various dates from 2008. *Coral Reefs of the World.* Springer, Dordrecht. An ongoing series of research level books by multiple authors. Most are based on a single region, some are based on themes:

1 *Coral Reefs of the USA.* Eds: Riegl, B., and Dodge, B. M. 2008.

2 *The Great Barrier Reef.* Eds: Hutchings, P., Kingsford, M. J., and Hoegh-Guldberg, O. 2009.

3 *The Gulf.* Eds: Riegl, B. M., and Purkis, S. J. 2016.

4 *United Kingdom Overseas Territories.* Ed.: Sheppard, C. R. C. 2013.

5 *Coral Reef Science.* Ed.: Kayanne, H. 2016.

6 *Coral Reefs at the Crossroads.* Eds: Hubbard, D. K. et al. 2016.

8 *Eastern Tropical Pacific.* Eds: Glynn, P. W., Manzella, D. P., and Wnochs, I. C. 2016.

11 *Red Sea.* Eds: Voolstra, C. R., and Berumen, M. L. 2019.

12 *Mesophotic Reefs.* Eds: Loya, Y., Puglise, K. A., and Bridge, T. C. L. 2019.

13 *Japan.* Eds: Iguchi, A., and Hongo, C. 2018.

Roberts, C. M. 2019. *Reef Life: An Underwater Memoir.* Profile Books, London, 366 pp.

Rower, F. 2010. *Coral Reefs in the Microbial Seas.* Plaid Press, San Francisco, 201 pp.

Sale, P. (Ed.). 2002. *Coral Reef Fishes: Dynamics and Diversity in a Complex Ecosystem.* Academic Press, London, 1653 pp.

Sheppard, A. L. S. 2015. *Coral Reefs: Secret Cities of the Seas.* Natural History Museum, London, 112 pp.

Sheppard, C. R. C. 1983. *A Natural History of the Coral Reef.* Blandford Press, Poole, 160 pp.

Sheppard, C. R. C., Davy, S., Pilling, G., and Graham, N. G. 2018. *Biology of Coral Reefs.* 2nd edition. Oxford University Press, Oxford, 339 pp.

Sheppard, C. R. C., Price, A. R. G., and Roberts, C. J. 1992. *Marine Ecology of the Arabian Area: Patterns and Processes in Extreme Tropical Environments*. Academic Press, London, 380 pp.

Spalding, M., Ravilious, C., and Green, E. P. 2001. *Atlas of Coral Reefs*. University of California Press, Berkeley and Los Angeles, 425 pp.

Veron, J. E. N. 1995. *Corals in Space and Time: The Biogeography and Evolution of the Scleractinia*. UNSW Press, Sydney, 321 pp.

Veron, J. E. N. 2000. *Corals of the World* (3 vols). Australian Inst. Marine Science, Townsville, 1350 pp.

Veron, J. E. N. 2007. *A Reef in Time: The Great Barrier Reef from Beginning to End*. Harvard University Press, Cambridge, Mass., 207 pp.

Wilkinson, C. R. 2008. *Status of Coral Reefs of the World: 2008*. Global Coral Reef Monitoring Network and Reef and Rainforest Research Centre, Townsville, 298 pp.

Woodley, C. M., et al. 2016. *Diseases of Coral*. Wiley, London, 582 pp.

Scientific reviews

Abesamis, R. A., Green, A. L., Russ, G. R., and Jadloc, R. L. 2014. The intrinsic vulnerability to fishing of coral reef fishes and their differential recovery in fishery closures. *Reviews in Fish Biol. and Fisheries* 24: 1033–63.

Anthony, K. R. N. 2016. Coral reefs under climate change and ocean acidification: challenges and opportunities for management and policy. *Annu. Rev. Environ. Resour.* 41: 59–81.

Ateweberhan, M., McClanahan, T. A., Graham, N. A. J., and Sheppard, C. R. C. 2011. Episodic heterogeneous decline and recovery of coral cover in the Indian Ocean. *Coral Reefs* 30: 739–52.

Bradbury, R. H., and Seymour, R. M. 2009. Coral reef science and the new commons. *Coral Reefs* 28: 831–7.

Bruno, J. F., and Selig, E. R. 2007. Regional decline of coral cover in the Indo-Pacific: timing, extent, and subregional comparisons. *PLoS ONE* 2 (8): e711.

Camp, E. F., Schoepf, V., Mumby, P. J., and Suggett, D. J. 2018. The future of coral reefs subject to rapid climate change: lessons from natural extreme environments. *Front. Mar. Sci.* 5: 433. (Editorial, prefacing 15 research studies.)

Carpenter, K., and 40 others 2008. One-third of reef-building corals face elevated extinction risk from climate change and local impacts. *Science* 321: 560–3.

Gardner, T. A., Côté, I. M., Gill, J. A., Grant, A., and Watkinson, A. R. 2003. Long-term regionwide declines in Caribbean corals. *Science* 301: 958–60.

Graham, N. A. J., and McClanahan, T. R. 2013. The last call for marine wilderness? *BioScience* 63: 397–402.

Green, A. L., Maypa, A. P., Almany, G. R., Rhodes, K. L., Weeks, R., Abesamis, R. A., Gleason, M. G., Mumby, P. J., and White, A. T. 2014. Larval dispersal and movement patterns of coral reef fishes, and implications for marine reserve network design. *Biological Reviews* 90: 1215–47.

Harvey, B. J., Nash, K. L., Blanchard, J. L., andEdwards, D. P. 2018. Ecosystem-based management of coral reefs under climate change. *Ecology and Evolution* 8: 6354–68.

Hoegh-Guldberg, O., and Bruno, J. F. 2010. The impact of climate change on the world's marine ecosystems. *Science* 328: 1523–8.

Hoegh-Guldberg, O., Mumby, P. J., Hooten, A. J., Steneck, R. S., et al., 2007. Coral reefs under rapid climate change and ocean acidification. *Science* 318: 1737–42.

Hughes, T. P., Graham, N. A. J., Jackson, J. B. C., Mumby, P. J., and Steneck, R. S. 2010. Rising to the challenge of sustaining coral reef resilience. *Trends Ecology and Evolution* 1282 (25): 633–42.

Jackson, J. B. C., Kirby, M. X., Berger, W. H., Bjorndal, K. A., et al., 2001. Historical overfishing and the recent collapse of coastal ecosystems. *Science* 293: 629–37.

Knowlton, N., and Jackson, J. B. C. 2008. Shifting baselines, local impacts, and global change on coral reefs. *PLoS Biology* 6 (2): e54.

McClanahan, T. R., and Omukoto, J. O. 2011. Comparison of modern and historical fish catches (AD 750–1400) to inform goals for marine protected areas and sustainable fisheries. *Conservation Biology* 25: 945–55.

Pandolfi, M. 2015. Incorporating uncertainty in predicting the future response of coral reefs to climate change. *Annu. Rev. Ecol. Evol. Syst.* 46: 281–303.

Sully, S., Burkepile, D. E., Donovan, M. K., Hodgson, G., and van Woesik, R. 2019. A global analysis of coral bleaching over the past two decades. DOI: <https://doi.org/10.1038/s41467-019-09238-2>.

Thurber, R. V., Payet, J. P., Thurber, A. R., and Correa, A. M. S. 2017. Virus–host interactions and their roles in coral reef health and disease. *Nature Reviews Microbiology* 15: 205–16.

Veron, J. E., Hoegh-Guldberg, O., Lenton, T. M., Lough, J. M., et al. 2009. The coral reef crisis: the critical importance of <350 ppm CO_2. *Marine Pollution Bulletin* 58: 1428–36.

Wilkinson, C., and Salvat, B. 2012. Coastal resource degradation in the tropics: does the tragedy of the commons apply for coral reefs, mangrove forests and seagrass beds. *Marine Pollution Bulletin* 64: 1096–105.

Index

For the benefit of digital users, indexed terms that span two pages (e.g., 52–53) may, on occasion, appear on only one of those pages.

index